POCKET ANATOMY
& PHYSIOLOGY

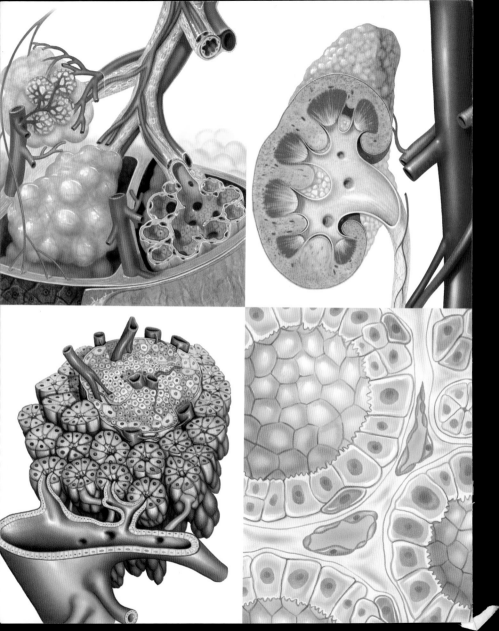

POCKET **ANATOMY**
& PHYSIOLOGY

The Compact Guide to the Human Body and How It Works

PROFESSOR KEN ASHWELL BMEDSC, MBBS, PHD

First edition for North America published in 2016 by Barron's Educational Series, Inc.

A Global Book
© 2016 Quarto Publishing PLC, 6 Blundell Street, London N7 9BH, UK

Conceived, designed, and produced by Global Book Publishing

Illustrations by Joanna Culley BA(Hons) RMIP, MMAA, IMI (Medical-Artist.com)

Additional illustrations: David Carroll, Peter Child, Deborah Clarke, Geoff Cook, Marcus Cremonese, Beth Croce, Wendy de Paauw, Levant Efe, Hans De Haas, Mike Golding, Mike Gorman, Jeff Lang, Alex Lavroff, Ulrich Lehmann, Ruth Lindsay, Richard McKenna, Annabel Milne, Tony Pyrzakowski, Oliver Rennert, Caroline Rodrigues, Otto Schmidinger, Bob Seal, Vicky Short, Graeme Tavendale, Jonathan Tidball, Paul Tresnan, Valentin Varetsa, Glen Vause, Spike Wademan, Trevor Weekes, Paul Williams, David Wood

All inquiries should be addressed to:
Barron's Educational Series, Inc.
250 Wireless Boulevard
Hauppauge, NY 11788
www.barronseduc.com

ISBN: 978-1-4380-0905-6

Library of Congress Control Number: 2016942330

Printed in China

9 8 7 6 5 4 3 2 1

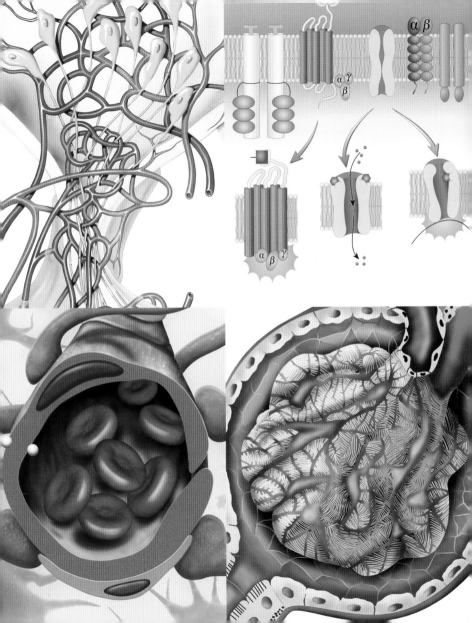

Contents

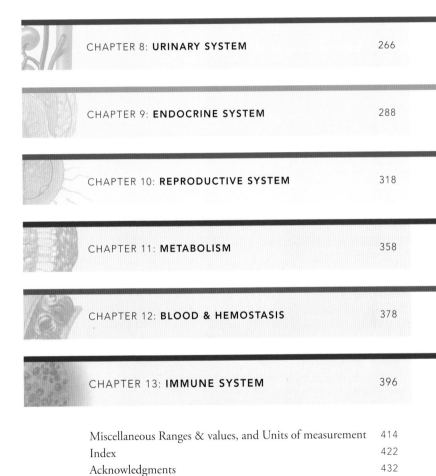

Introduction

ANATOMY AND PHYSIOLOGY ARE THE TWO KEY BIOLOGICAL DISCIPLINES THAT UNDERPIN STUDIES IN ALL HEALTHCARE PROFESSIONS. Anatomy is the study of the structure of the human body at all levels from the naked eye to the microscopic, whereas physiology is the study of the function of the body based on the application of the principles of chemical and physical science to living tissue. Biochemistry is simply structure and function taken to the molecular level and so is an extension of physiology.

Structure and function must be considered and understood together, because they are intimately entwined, so mastery of both is essential for any studies in medical science. Knowledge of structure is meaningless without an understanding of the accompanying function, and it is impossible to understand function without knowledge of the relevant structure. Furthermore, evidence-based clinical practice is built on a detailed understanding of body structure and function and its application in the clinic. Without a sound knowledge of anatomy and physiology it is not only impossible to understand how disease processes affect body function, but it is also very difficult to understand the biological basis of clinical symptoms and signs. Students who do not understand the underlying biology of a patient's clinical condition are simply rote learning patterns to make diagnoses and will not be able to apply their knowledge flexibly in different clinical conditions. All health practitioners and medical educators know that intelligent clinical practice can only be achieved through a sound understanding of the basic

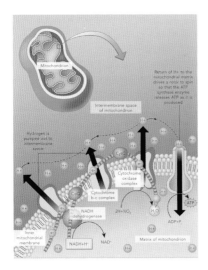

Each topic has succinct yet informative text linked to clear illustrations of the relevant body part or flow charts to explain difficult physiological concepts. The aim is to provide a compact, yet detailed, summary of the most important topics in human structure and function. Useful reference data on normal laboratory findings have also been provided so students can learn typical values and apply these in a clinical setting when interpreting the results of patients' tests.

The book is organized by body systems for easy reference, with a comprehensive index for easy location of the key themes and concepts. It can be read as a sequence of themes or used as a reference book for specific topics. It offers a handy yet comprehensive reference for all students of nursing, chiropractic, occupational therapy, physiotherapy, medicine, and dentistry, and anyone with an interest in human biology and systems.

sciences and a keen appreciation of how altered structure leads to changes in function.

In this book we have brought together key topics in the structure and function of the human body and presented them in an easy-to-understand format that students can access readily. Themes have been chosen carefully to provide a balanced coverage of the most important ideas in each body system, ranging from the cellular level up to the macroscopic.

TOP LEFT *Aerobic metabolism*
OPPOSITE *Neuromuscular junction*

Homeostasis: keeping the balance

A KEY PRINCIPLE IN PHYSIOLOGY IS THAT OF HOMEOSTASIS, or internal balance. This is the tendency for living things to maintain a constant internal environment, despite changes in the external environment. An example would be the maintenance of the human core temperature at around 98.6°F (37.0°C) despite a fluctuating environmental temperature.

Other internal physiological variables that must be kept constant include metabolic rate, blood pressure, arterial oxygen concentration, arterial carbon dioxide concentration, blood glucose concentration, blood pH, and concentrations of various ions in the blood (e.g., sodium, potassium, calcium, and phosphate). Each of these has to be maintained within a narrow range around a set point that is optimal for survival and reproductive success.

Control systems that maintain homeostasis require three components: a sensory receptor that detects the variable, an integrator or control center that decides whether the variable has deviated from the set point, and an effector that brings about an internal

▼ THE KEY ELEMENTS OF A HOMEOSTATIC CONTROL SYSTEM

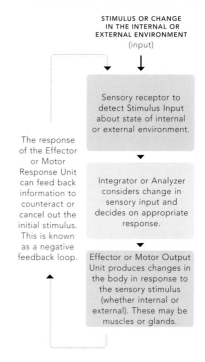

STIMULUS OR CHANGE IN THE INTERNAL OR EXTERNAL ENVIRONMENT
(input)

Sensory receptor to detect Stimulus Input about state of internal or external environment.

The response of the Effector or Motor Response Unit can feed back information to counteract or cancel out the initial stimulus. This is known as a negative feedback loop.

Integrator or Analyzer considers change in sensory input and decides on appropriate response.

Effector or Motor Output Unit produces changes in the body in response to the sensory stimulus (whether internal or external). These may be muscles or glands.

change that returns the variable to the set point.

The combination of these elements produces negative feedback loops. Control of some variables, such as blood calcium concentration,

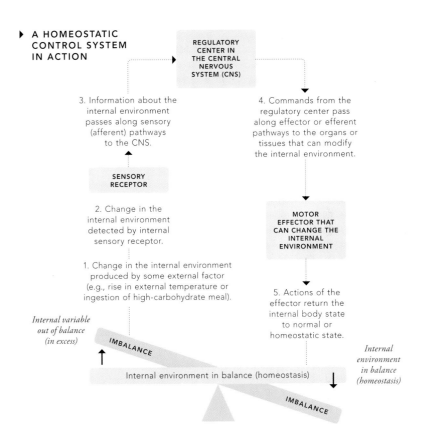

▶ A HOMEOSTATIC CONTROL SYSTEM IN ACTION

REGULATORY CENTER IN THE CENTRAL NERVOUS SYSTEM (CNS)

3. Information about the internal environment passes along sensory (afferent) pathways to the CNS.

4. Commands from the regulatory center pass along effector or efferent pathways to the organs or tissues that can modify the internal environment.

SENSORY RECEPTOR

2. Change in the internal environment detected by internal sensory receptor.

MOTOR EFFECTOR THAT CAN CHANGE THE INTERNAL ENVIRONMENT

1. Change in the internal environment produced by some external factor (e.g., rise in external temperature or ingestion of high-carbohydrate meal).

5. Actions of the effector return the internal body state to normal or homeostatic state.

Internal variable out of balance (in excess)

IMBALANCE

Internal environment in balance (homeostasis)

Internal environment in balance (homeostasis)

IMBALANCE

may require two feedback loops acting in concert but detecting different changes in the variable—for example, either a rise or a fall in the calcium concentration. Different hormones then act in response to the feedback, e.g., the parathyroid hormone mediates the response to lowered blood calcium, while calcitonin mediates the response to elevated blood calcium.

Nervous system

THE NERVOUS SYSTEM IS RESPONSIBLE FOR DETECTING ANY CHANGES IN THE INTERNAL AND EXTERNAL ENVIRONMENT (sensory organs), making decisions on how to respond to those changes (integrative nervous connections in the brain and spinal cord) and producing changes in the internal and external environment by acting on either internal organs, such as glands or smooth muscles of the gut tube, or the voluntary skeletal muscles (effector components).

Both the nervous and endocrine systems are critically important for maintaining homeostasis, but they act over different time frames. The nervous system tends to respond to changes in the environment within seconds to minutes, whereas the endocrine system responds over hours to years. There are also several higher functions that are part of the role of the nervous system, including sleep, emotional responses, learning and memory, social interaction, language, musical appreciation, and cognitive performance.

The nervous system is divided into the central nervous system (CNS), which consists of the brain and spinal cord, and the peripheral nervous system (PNS), which consists of peripheral nerves and collections of nerve cell bodies called ganglia. Ganglia in the PNS may be sensory or effector, i.e., controlling glands or smooth muscle. Both central and peripheral parts of the nervous system are made up of nerve cells, which transmit and process information, and a variety of support cells, such as glia in the CNS and Schwann cells in the PNS. The CNS has a rich blood supply because of its high metabolic rate.

▶ **THE MAJOR PARTS OF THE CENTRAL AND PERIPHERAL NERVOUS SYSTEMS**

The central nervous system consists of the brain and spinal cord. The spinal cord has enlargements for control of the upper and lower limbs (cervical and lumbosacral enlargements) and ends at the lower back just beyond the level of the lowest rib. The collected lumbar and sacral nerves below that point form a structure called the *cauda equina* (Latin for "horse's tail"). The peripheral nervous system consist of peripheral nerves that form networks, e.g., the brachial and lumbosacral plexuses, to supply the limbs.

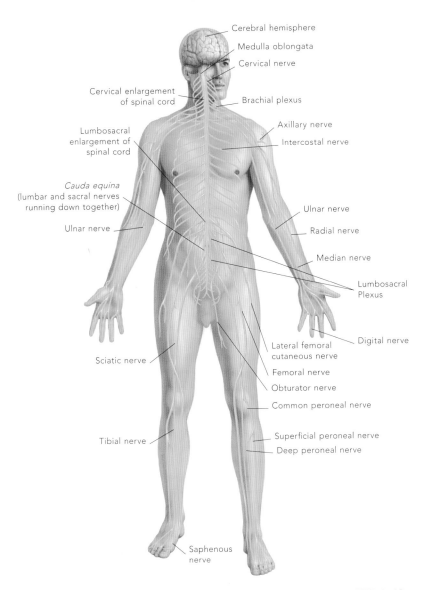

Cerebral hemisphere

Medulla oblongata

Cervical nerve

Cervical enlargement of spinal cord

Brachial plexus

Axillary nerve

Intercostal nerve

Lumbosacral enlargement of spinal cord

Cauda equina (lumbar and sacral nerves running down together)

Ulnar nerve

Ulnar nerve

Radial nerve

Median nerve

Lumbosacral Plexus

Digital nerve

Sciatic nerve

Lateral femoral cutaneous nerve

Femoral nerve

Obturator nerve

Common peroneal nerve

Superficial peroneal nerve

Deep peroneal nerve

Tibial nerve

Saphenous nerve

Musculoskeletal system

THE MUSCULOSKELETAL SYSTEM IS RESPONSIBLE FOR PRODUCING VOLUNTARY MOVEMENT (i.e., movement that the individual consciously generates), providing structure to the body, and protecting soft and vulnerable internal organs. The key components of the musculoskeletal system are: bones that provide protection for internal organs and a framework for attachment of the muscles, joints that allow bones to be joined together and (where appropriate) allow movement of bones relative to each other, and the voluntary or skeletal muscles that produce body movement.

The bones of the musculoskeletal system must constantly remodel during life in response to physical stresses and growth. Muscles are also in a constant state of flux as they respond to increased workloads from exercise.

There are various important types of tissue within the musculoskeletal system, including articular or hyaline cartilage that provides a low friction surface at mobile synovial joints and dense connective tissue that provides attachments of bones to bones (ligaments) or muscles to bones (tendons). Other connective tissue sheets separate groups of skeletal muscles, protect vessels and nerves traversing the spaces between muscle groups, and provide connective tissue sheaths that enhance muscle function.

The musculoskeletal system is closely integrated with other body systems. For example, movement must be controlled by the nervous system. The bones themselves provide an important store for calcium and phosphate and contribute to the maintenance of an acid/base balance in the body's fluids. Some bones house the hematopoietic tissue that produces red and white blood cells and stores important fat reserves.

▶ **SKELETAL MUSCLES**

The voluntary or skeletal muscles of the body have at least one attachment to the skeleton and move under voluntary command from the brain and spinal cord. For each muscle, the attachment closest to the body's midline is called the origin and the attachment that is further away is called the insertion.

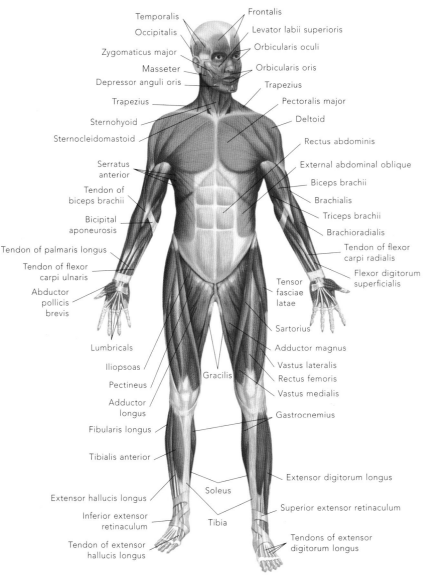

Temporalis

Occipitalis

Zygomaticus major

Masseter

Depressor anguli oris

Trapezius

Sternohyoid

Sternocleidomastoid

Serratus anterior

Tendon of biceps brachii

Bicipital aponeurosis

Tendon of palmaris longus

Tendon of flexor carpi ulnaris

Abductor pollicis brevis

Lumbricals

Iliopsoas

Pectineus

Adductor longus

Fibularis longus

Tibialis anterior

Extensor hallucis longus

Inferior extensor retinaculum

Tendon of extensor hallucis longus

Frontalis

Levator labii superioris

Orbicularis oculi

Orbicularis oris

Trapezius

Pectoralis major

Deltoid

Rectus abdominis

External abdominal oblique

Biceps brachii

Brachialis

Triceps brachii

Brachioradialis

Tendon of flexor carpi radialis

Flexor digitorum superficialis

Tensor fasciae latae

Sartorius

Adductor magnus

Vastus lateralis

Rectus femoris

Vastus medialis

Gastrocnemius

Extensor digitorum longus

Superior extensor retinaculum

Tendons of extensor digitorum longus

Gracilis

Soleus

Tibia

Circulatory system

THE CIRCULATORY SYSTEM SERVES TO MOVE NUTRIENTS, hormones, essential gases, and waste products around the body. It also contributes to immune function by transporting white blood cells and immune system proteins like antibodies and complement to sites of infection or cancer growth. The circulatory system may also play a part in the maintenance of a constant body temperature by shifting heat from the body core to the skin surface.

The circulatory system consists of a pump (the heart) and two major components: the pulmonary and systemic circulations. The pulmonary circulation carries deoxygenated blood from the right side of the heart to the lungs by the pulmonary arterial branches, provides a capillary bed for gas exchange with inhaled air in the lungs, and returns the oxygenated blood to the left side of the heart through the pulmonary venous channels. The systemic circulation carries oxygenated blood from the left side of the heart to all the other organs of the body through the aorta and its branches, provides a capillary bed for gas exchange at the tissues, and returns the deoxygenated venous blood to the right side of the heart through the systemic veins.

The amount of fluid reaching the tissues through systemic arteries is slightly in excess of that returning through systemic veins, so the excess fluid in the peripheral tissues must be returned to the venous circulation through lymphatic channels. These channels provide an opportunity to check for foreign invaders such as bacteria and viruses, and mutant cells, e.g., cancer cells.

The heart consists of involuntary striated (cardiac) muscle that must beat rhythmically from early in embryonic life to the moment of death. Cardiac muscle has important functional attributes—intrinsic rhythmicity and electrical connections between cells—to ensure that rhythmic cardiac contraction can be sustained.

▶ **ARTERIES AND VEINS OF THE SYSTEMIC CIRCULATION**
Arteries (colored red) carry oxygenated blood away from the left ventricle of the heart to the capillary beds of the tissues and veins (colored blue), return deoxygenated blood to the right atrium.

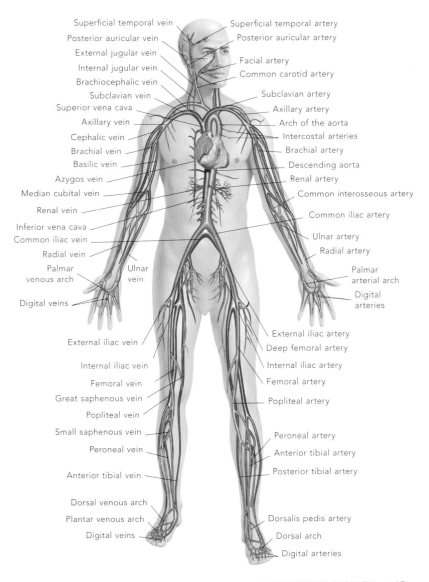

Superficial temporal vein
Posterior auricular vein
External jugular vein
Internal jugular vein
Brachiocephalic vein
Subclavian vein
Superior vena cava
Axillary vein
Cephalic vein
Brachial vein
Basilic vein
Azygos vein
Median cubital vein
Renal vein
Inferior vena cava
Common iliac vein
Radial vein
Palmar venous arch
Ulnar vein
Digital veins

Superficial temporal artery
Posterior auricular artery
Facial artery
Common carotid artery
Subclavian artery
Axillary artery
Arch of the aorta
Intercostal arteries
Brachial artery
Descending aorta
Renal artery
Common interosseous artery
Common iliac artery
Ulnar artery
Radial artery
Palmar arterial arch
Digital arteries

External iliac vein
Internal iliac vein
Femoral vein
Great saphenous vein
Popliteal vein
Small saphenous vein
Peroneal vein
Anterior tibial vein
Dorsal venous arch
Plantar venous arch
Digital veins

External iliac artery
Deep femoral artery
Internal iliac artery
Femoral artery
Popliteal artery
Peroneal artery
Anterior tibial artery
Posterior tibial artery
Dorsalis pedis artery
Dorsal arch
Digital arteries

Respiratory system

THE RESPIRATORY SYSTEM IS CONCERNED WITH BRINGING OXYGEN FROM THE EXTERNAL ENVIRONMENT INTO THE BODY, providing gas exchange with the blood, and returning carbon dioxide to the external environment. It also has some accessory functions, such as providing a means for sensory sampling of the inhaled air (the sensory nose), assisting with maintaining the pH of the blood (through regulation of carbon dioxide concentration in the blood), and providing a source of exhaled air for producing the voice (phonation).

The respiratory system consists of the external nose, nasal cavity, nasopharynx, and larynx (sometimes grouped as the upper respiratory tract), and the trachea, main bronchi, and lungs (often grouped as the lower respiratory tract). There are important accessory structures that contribute to respiratory function. These include the 12 ribs and respiratory muscles (intercostal muscles and muscular diaphragm) that ventilate the lungs and the pleural sacs that enclose each lung to allow free movement of the lung during expansion.

Breathing or ventilation also requires a sophisticated control system to regulate the depth and rate of lung ventilation in response to changes in concentration of oxygen and carbon dioxide in the arterial blood. This involves sensory receptors in the head arteries or brainstem tissue, control and regulatory centers in the hypothalamus and brainstem, and nerves from the spinal cord to the respiratory muscles. The ongoing exposure of respiratory tissue to the external environment makes it vulnerable to infection and diseases caused by dust.

▸ **MAJOR PARTS OF THE RESPIRATORY SYSTEM**

The respiratory system consists of a conduction part that brings air into the lungs (nasal cavity, upper pharynx, larynx, trachea, and bronchi) and an exchange part at the level of the alveoli. The trachea branches into right and left primary or main bronchi, which in turn divide into lobar bronchi (three in the right lung, two in the left).

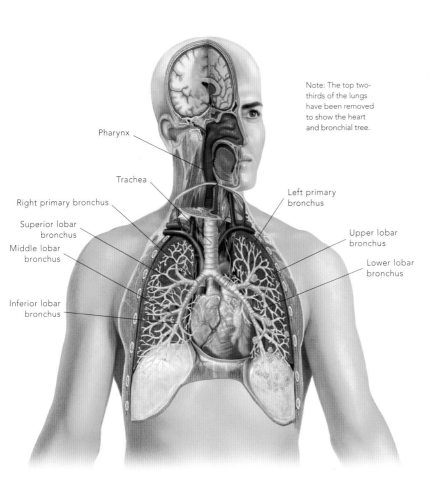

Pharynx

Trachea

Right primary bronchus

Superior lobar bronchus

Middle lobar bronchus

Inferior lobar bronchus

Note: The top two-thirds of the lungs have been removed to show the heart and bronchial tree.

Left primary bronchus

Upper lobar bronchus

Lower lobar bronchus

Digestive system

THE DIGESTIVE SYSTEM IS RESPONSIBLE FOR THE INGESTION OF FOOD AND WATER; the mechanical, chemical, and enzymatic processing of that food to reduce it to its biochemical constituents (digestion in its true sense); the absorption of useful nutrients from the digested food; and finally the excretion of waste products to the external environment.

The digestive system is essentially an elongated and convoluted tube, but with important accessory glands and sacs attached as side structures. The tubular part of the digestive system consists of the mouth or oral cavity, tongue and teeth, oropharynx and laryngopharynx, esophagus, stomach, small and large intestines, and the anus. The accessory glands and structures are the major salivary glands around the oral cavity, and the liver, gall bladder, and pancreas, which are developmental side growths of the upper small intestine. The digestive and respiratory systems cross at the oropharynx, which can cause problems with choking when food and water mistakenly enter the respiratory tract.

The digestive system, like the respiratory system, is a portal for foreign invaders to enter the body and cause infection. For this reason, clusters of immune system cells (the tonsils) are located at the entrance to the digestive system. The lower gut tube is also home to many friendly microorganisms that actually help digestion and nutrient supply.

The invasion of the body by ingested disease microorganisms or usually friendly gut bacteria is prevented by abundant immune system cells (mucosa-associated lymphoid tissue) in the wall of the gut tube. The liver is also critically important in detoxifying those chemicals that have been produced by microbes and absorbed through the gut wall.

▶ **MAJOR ORGANS OF THE DIGESTIVE TRACT**

Most of the digestive system is located in the abdominal cavity, where the upper third of the abdomen houses the stomach, liver, gall bladder, and pancreas, and the lower two-thirds of the abdomen accommodate lengths of small and large intestine.

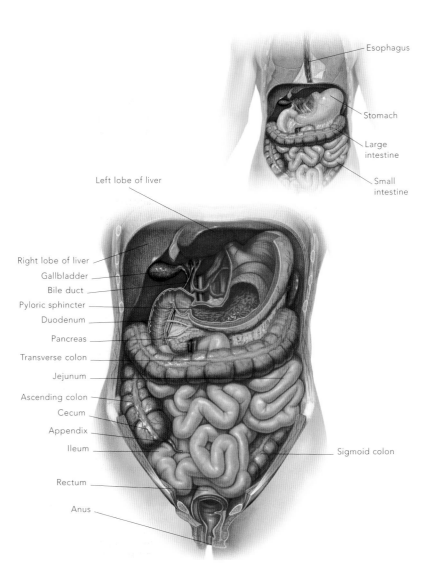

Esophagus

Stomach

Large intestine

Small intestine

Left lobe of liver

Right lobe of liver

Gallbladder

Bile duct

Pyloric sphincter

Duodenum

Pancreas

Transverse colon

Jejunum

Ascending colon

Cecum

Appendix

Ileum

Rectum

Anus

Sigmoid colon

Urinary system

THE URINARY SYSTEM IS CONCERNED WITH THE ULTRAFILTRATION OF THE BLOOD TO SEPARATE PLASMA PROTEINS FROM DISSOLVED SUBSTANCES; the reabsorption of important ions and nutrients—e.g., sodium, potassium, calcium, glucose, and amino acids—from the ultrafiltrate; and the excretion of nitrogen-containing waste (mainly as the compound urea), other toxins, and some drugs to the external environment. This system also plays critical roles in regulating blood pressure, pH of the blood, water balance, activation of vitamin D, and production of red blood cells.

The macroscopic or naked eye components of the urinary system consist of the paired kidneys, the tubular ureters to carry urine to the urinary bladder for storage, and the urethra to carry urine to the external environment.

The internal structure of the kidney is very complex, and each kidney contains more than one million functional units called nephrons. Each nephron has parts concerned with ultrafiltration (the glomeruli) as well as a complex series of tubules concerned with reabsorption and/or secretion of important substances.

The lower parts of the urinary system are closely linked with the reproductive system, particularly in males, where the urethra provides a common pathway for both urine excretion and ejaculation of semen.

▶ **ORGANS OF THE URINARY SYSTEM**

The urinary system consists of paired kidneys located in the middle back immediately in front of the 12th ribs, long tubular ureters to carry urine to the urinary bladder, and the urethra to allow urine to exit the body. The urinary bladder is a pelvic organ unless it is very full, in which case it may rise above the pubic symphysis. The urethra is short in females (about 1.5 inches/4 cm long) and long in males (about 8 inches/20 cm).

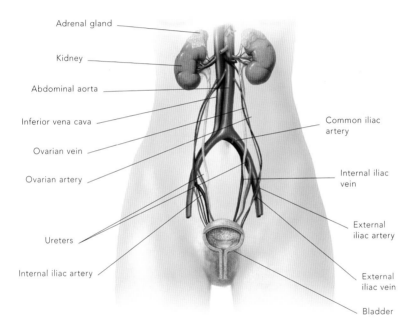

Adrenal gland

Kidney

Abdominal aorta

Inferior vena cava

Ovarian vein

Ovarian artery

Ureters

Internal iliac artery

Common iliac artery

Internal iliac vein

External iliac artery

External iliac vein

Bladder

Reproductive system

THE REPRODUCTIVE SYSTEM PROVIDES FOR THE PRODUCTION AND NOURISHMENT OF THE NEXT GENERATION. The system consists of paired gonads in each sex—testes in males and ovaries in females. Gonads perform endocrine functions and produce the sex cells (ovum or egg in females, sperm cells in males). In females the ovaries are internal, but in males the testes sit outside the body to provide an optimal temperature for sperm cell production that is a few degrees below core temperature.

In females there are paired uterine or Fallopian tubes that carry the eggs down toward the uterine body, where a fertilized egg can implant to develop into an embryo and fetus. The uterus is also the organ of parturition—the process of giving birth—and has smooth muscle that is capable of rhythmic contraction during labor to expel the fetus and placenta through the vagina. Finally, there are female external genitalia around the opening of the urethra and vagina. Women also possess breast tissue for lactation to support the newborn.

In males there is an accessory structure adjacent to the testis known as the epididymis, which supports maturing sperm; a ductus or vas deferens to carry sperm from the epididymis to the rest of the reproductive tract; accessory glands (seminal vesicles and prostate); and the urethra, down which semen (the sperm cells plus the products of the male accessory glands) can be ejaculated into the female reproductive tract. The male urethra traverses an organ of erectile tissue called the penis, which can stiffen to allow penetration of the vagina during sexual intercourse.

▶ **ORGANS OF THE FEMALE AND MALE REPRODUCTIVE SYSTEMS**

The female reproductive system consists of ovaries to produce eggs and hormones; the Fallopian or uterine tubes to carry the ovum toward the uterus; and the uterus, an organ of gestation and parturition. The male reproductive system has external gonads (the testes), male accessory organs, and an organ of sexual intercourse (the penis).

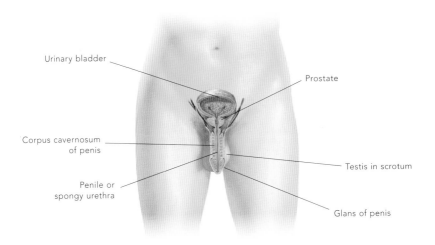

Urinary bladder

Prostate

Corpus cavernosum
of penis

Testis in scrotum

Penile or
spongy urethra

Glans of penis

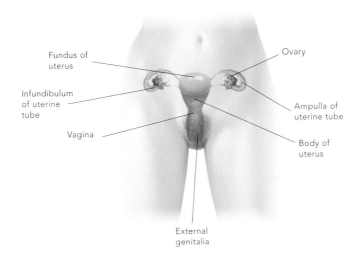

Fundus of
uterus

Ovary

Infundibulum
of uterine
tube

Ampulla of
uterine tube

Vagina

Body of
uterus

External
genitalia

Endocrine system

THE ENDOCRINE SYSTEM CONSISTS OF A GROUP OF DUCTLESS GLANDS DISTRIBUTED THROUGHOUT THE BODY. Like the nervous system, they function to maintain a constant internal environment, but they do so over periods of hours to years. The endocrine system also plays a key role in growth and development. The nervous and endocrine systems are linked through a connection between the hypothalamus of the brain and the pituitary gland, the so-called master gland of the endocrine system that sits immediately below the brain.

Glands of the endocrine system secrete their products (either peptide or steroid hormones) mainly into the bloodstream, but also locally or into other body cavities. The pituitary gland secretes many hormones —e.g., thyroid stimulating hormone, adrenocorticotropic hormone (ACTH), oxytocin, and antidiuretic hormone—that act on other glands of the endocrine system or directly regulate growth, water conservation, parturition, or milk ejection from the breast. The thyroid gland regulates metabolic rate and contains C cells that participate in calcium homeostasis. The parathyroid glands also participate in calcium metabolism. The pancreas contains an endocrine part concerned with carbohydrate metabolism through secretion of insulin and glucagon into the bloodstream.

The adrenal glands consist of an outer part—the adrenal cortex—that secretes steroid hormones and an inner part— the adrenal medulla—that secretes epinephrine and norepinephrine. The testes and ovaries are also endocrine glands that produce sex steroids. The placenta plays an endocrine role during pregnancy, producing hormones that stimulate breast development and loosen the pelvic ligaments to make birth easier.

▶ **ORGANS OF THE ENDOCRINE SYSTEM**

The endocrine system consists of ductless glands that produce peptide or steroid hormones to regulate body metabolism. Many endocrine glands —e.g., thyroid, adrenal cortex, ovaries, and testes—are regulated by the pituitary, the master gland of the endocrine system.

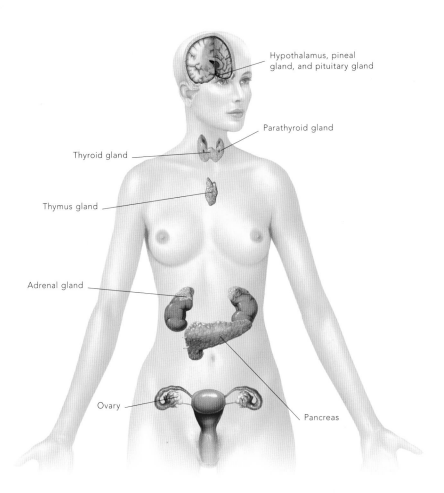

Hypothalamus, pineal gland, and pituitary gland

Parathyroid gland

Thyroid gland

Thymus gland

Adrenal gland

Ovary

Pancreas

Blood

BLOOD IS A SPECIALIZED TYPE OF
CONNECTIVE TISSUE THAT TRANSPORTS
GASES, nutrients, waste products,
immune cells, proteins, vitamins, and
hormones around the body. Blood
consists of a fluid component (plasma)
with dissolved substances, e.g., sugars,
amino acids, hormones, urea, and
electrolytes; a suspended plasma
protein component, e.g., albumen,
immunoglobulins, clotting factors,
and transport molecules; and a cellular
component, i.e., red and white blood
cells and platelets.

Red blood cells contain hemoglobin
for transport of oxygen, whereas white
blood cells are part of the immune
system, defending the body against
either foreign invaders, such as
bacteria, parasites, rickettsiae, fungi,
and viruses or mutant body cells, such
as cancers. Platelets play a critical role
in blood clotting. The red color of
blood is due to the presence of a red
pigment (heme) in hemoglobin, the
molecular complex responsible for
carrying oxygen in red blood cells.

Blood is pumped around the body
by the circulatory system, but its
composition changes slightly
depending on the part of the
circulatory system that is being
traversed. This is because the fluid
component leaks into tissue spaces
around the arterial end of the capillary
bed to be mainly reabsorbed before
the capillaries rejoin to form venules.
Blood also assists in transporting heat
between different parts of the body
to contribute to thermoregulation.

▶ A TYPICAL CAPILLARY

Blood is a type of connective tissue and
contains cells (red blood cells, platelets,
and leukocytes) suspended in a fluid
matrix (plasma) with dissolved proteins
(plasma proteins like albumen and
immunoglobulins). At the capillary bed
(shown here), there is interchange across
the capillary wall of some of the fluid
component of the blood with the
extravascular fluid around the capillary.

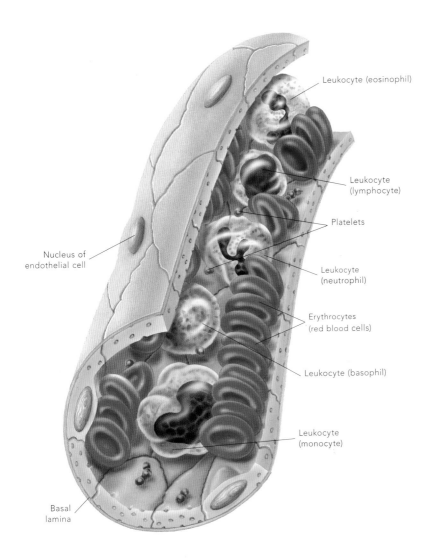

Leukocyte (eosinophil)

Leukocyte (lymphocyte)

Platelets

Leukocyte (neutrophil)

Nucleus of endothelial cell

Erythrocytes (red blood cells)

Leukocyte (basophil)

Leukocyte (monocyte)

Basal lamina

Integumentary system

THE INTEGUMENTARY SYSTEM, OR SKIN, IS THE SURFACE LAYER OF THE BODY. As such, it is the system with the largest surface area and makes up between 10% and 15% of body weight. The skin not only provides protection against the physical, chemical, and biological hazards of the external environment, but it also plays important roles in water conservation, sensation (touch, pain, vibration, and temperature), thermoregulation, vitamin D production, social communication (e.g., smiling and frowning), and even some excretion (lactic acid, urea, and some metals).

The skin consists of an external epithelial layer called the epidermis and a deeper connective tissue layer called the dermis over the hypodermis. The epidermis is continuously exposed to the external environment, and its cells must be replenished by continuous cell division at the base of the epidermis. As epidermal cells mature, they become increasingly filled with a protein called keratin until they are sloughed off as keratin flakes at the skin surface. Most skin overlies muscle and is separated from muscle tissue by connective tissue layers called deep fascia.

Specialized extensions of the epidermal tissue into the dermis form the sweat and sebaceous glands that respectively produce sweat for thermoregulation and sebum as an antimicrobial agent and epidermal protectant. Hairs are specialized keratin-containing products of the epidermis that assist in thermoregulation and, when in thick layers, may provide physical protection against abrasion and solar radiation. Nails are also specialized products of the epidermal nailbed and, like hairs, are composed of keratin.

▸ **LAYERS OF THE SKIN**

Skin may be hairy (as shown here) or thick—as seen on the palms or soles. Skin includes two types of glands (sweat and sebaceous), many different types of sensory receptors (e.g., Ruffini, Krause bulb, Meissner's, and Pacinian corpuscles), and a rich vascular supply. The epidermis contains several layers: stratum germinativum, stratum spinosum, stratum granulosum, and stratum corneum.

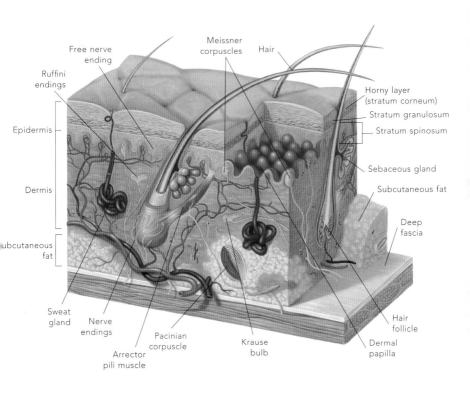

Free nerve ending

Ruffini endings

Meissner corpuscles

Hair

Epidermis

Horny layer (stratum corneum)

Stratum granulosum

Stratum spinosum

Dermis

Sebaceous gland

Subcutaneous fat

Deep fascia

Subcutaneous fat

Sweat gland

Nerve endings

Pacinian corpuscle

Krause bulb

Hair follicle

Dermal papilla

Arrector pili muscle

Immune system

THE IMMUNE SYSTEM IS CONCERNED WITH THE BODY'S RESPONSE TO INJURY, and protecting the body against foreign invaders, such as bacteria, parasites, rickettsiae, fungi, and viruses, and internal mutants, such as cancer cells. It is an anatomically diffuse system, with some components grouped as organs visible to the naked eye— e.g., the tonsils, thymus, spleen, and lymph nodes—but many of its constituent cells distributed throughout the body in blood or tissue spaces.

Some immune system components respond the same way to all disease-causing organisms (innate immunity), whereas other components respond very specifically to particular invaders (adaptive or specific immunity). Innate immunity relies on cells, such as granulocytes and macrophages, and molecules, such as complement and cytokines.

The specific immune system is broadly divided into a humoral immune system and a cell-mediated immune system on the basis of the types of responses it can mount. The former consists of antibodies and the cells that produce them, i.e., B lymphocytes and plasma cells. Antibodies clump to foreign proteins and microbes and either make them easier to engulf or block their harmful effects.

Cell-mediated immunity consists of different classes of T lymphocyte cells—e.g., helper and killer—and their cell-killer functions. Helper T cells can stimulate other immune responses, whereas cytotoxic or killer T cells destroy cells that have been infected by viruses or bacteria or that have undergone malignant change to become cancers.

▶ **LYMPH NODES AND LYMPHATIC CHANNELS**

Lymph nodes are clusters of immune system cells located at key sites along the lymphatic channels that drain excess tissue fluid. Major groups of lymph nodes are located at the main joints of the limbs, e.g., cubital and popliteal nodes at the elbow and knee, and at the bases of the limbs in the axilla and inguinal region, but most lymph nodes are located in the thoracic and abdominal cavities. The thoracic duct is the largest lymph channel of the body and drains all the lower body and the left side of the upper body.

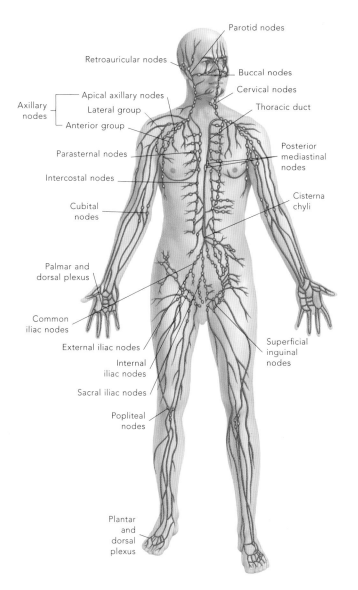

Parotid nodes

Retroauricular nodes

Buccal nodes

Cervical nodes

Apical axillary nodes

Thoracic duct

Axillary
nodes

Lateral group

Anterior group

Parasternal nodes

Posterior
mediastinal
nodes

Intercostal nodes

Cisterna
chyli

Cubital
nodes

Palmar and
dorsal plexus

Common
iliac nodes

External iliac nodes

Superficial
inguinal
nodes

Internal
iliac nodes

Sacral iliac nodes

Popliteal
nodes

Plantar
and
dorsal
plexus

Cell structure & organelles

THE CELL IS THE BASIC UNIT OF COMPLEX LIVING THINGS, AND THE HUMAN BODY CONSISTS OF CELLS AND THEIR EXTRACELLULAR PRODUCTS. The basic components of the cell are: i) a plasma membrane; surrounding ii) the cytoplasm, consisting of a fluid component called cytosol, with embedded organelles; and iii) a nucleus containing the DNA (deoxyribonucleic acid) that encodes genetic information and directs the metabolic actions of the cell. The plasma membrane plays a key role in regulating the entry of chemicals into the cell and has receptors involved in cell signaling.

The organelles include mitochondria, which use oxidative phosphorylation to produce energy for the cell in the form of adenosine triphosphate (ATP) and perform other nonoxidative metabolic functions; peroxisomes for breaking down fats and toxic substances; rough and smooth endoplasmic reticulum that produce protein and perform fat synthesis, respectively; the Golgi apparatus for packaging of materials for transport; and lysosomes for digestion of unwanted organelles.

All of these organelles are supported within a cytoskeleton framework made up of actin and intermediate filaments. The cytoskeleton also contributes to cellular extensions like microvilli and cilia that are found in some specialized cell types that line the interior of the digestive or respiratory tracts, respectively. The nucleus is surrounded by a nuclear membrane, with nuclear pores that regulate movement of messenger molecules into and out of the nucleus. The nucleus usually contains a region of condensed material known as the nucleolus, where ribosomes for protein manufacture are assembled.

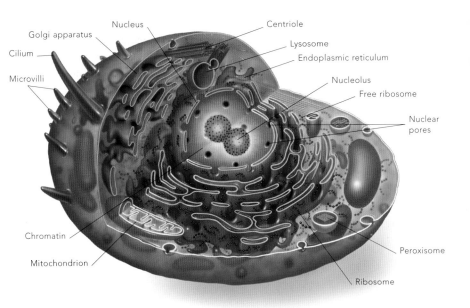

Golgi apparatus
Nucleus
Centriole
Cilium
Lysosome
Endoplasmic reticulum
Microvilli
Nucleolus
Free ribosome
Nuclear pores
Chromatin
Peroxisome
Mitochondrion
Ribosome

▲ **BASIC STRUCTURE OF THE CELL**

The cell includes many organelles with specific functions. These include: ribosomes for the manufacture of protein; endoplasmic reticulum for lipid synthesis; peroxisomes for breaking down fats; the Golgi apparatus for packaging of manufactured products; microvilli to increase the surface area of the cell for absorption, e.g., epithelial cells lining the wall of the small intestine; and cilia to move mucus or foreign material on the cell surface, e.g., epithelial cells lining the respiratory tract.

Chromosomes & genes

HUMAN DNA CONTAINS APPROXIMATELY 25,000 GENES AND IS GROUPED INTO SPECIALIZED STRUCTURES CALLED CHROMOSOMES. Every person has 22 pairs of non-sex chromosomes, plus an X and Y chromosome for a male or a pair of X chromosomes for a female. Deleted chromosomes, excess chromosomes, or fragments of chromosomes are usually associated with disease or disability, e.g., Down's syndrome.

Chromosomes consist of tightly coiled DNA strands and accompanying proteins but are only really visible when the chromosome threads are wound up immediately before cell division (see pp. 38–39). Double strands of DNA are wound around accompanying nuclear proteins called histones to produce beads called nucleosomes. Within each DNA strand, chains of nucleotides form genes, which are segments of DNA that usually specify the structure of a protein (see pp. 44–45). There are only four types of nucleotides in DNA (C—cytidine, A—adenosine, G—guanidine, and T—thymidine), and coding for each amino acid in the protein must rely on a combination of three nucleotides in a triplet, e.g., CAG. The average gene is 8,000 nucleotides long.

Some parts of each gene (exons) directly code for the amino acids of proteins, whereas other parts (introns) do not. Introns are not just junk DNA because they play a role in the evolution of new proteins and may help regulate gene expression. Changes in DNA are called mutations and can be due to faulty copying of DNA during cell division or the action of agents known as mutagens, e.g., ionizing radiation and mutagenic chemicals.

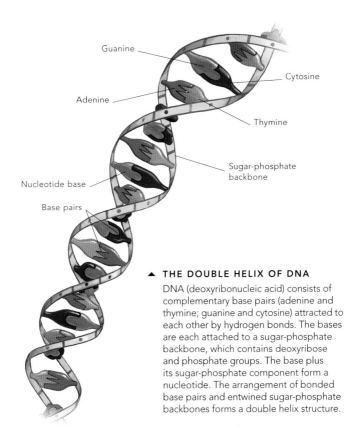

Guanine

Cytosine

Adenine

Thymine

Sugar-phosphate backbone

Nucleotide base

Base pairs

▲ THE DOUBLE HELIX OF DNA

DNA (deoxyribonucleic acid) consists of complementary base pairs (adenine and thymine; guanine and cytosine) attracted to each other by hydrogen bonds. The bases are each attached to a sugar-phosphate backbone, which contains deoxyribose and phosphate groups. The base plus its sugar-phosphate component form a nucleotide. The arrangement of bonded base pairs and entwined sugar-phosphate backbones forms a double helix structure.

DOWN'S SYNDROME

Down's syndrome is a chromosomal abnormality due to either the presence of three chromosome 21s in each cell (trisomy 21) or the translocation of parts of chromosome 21 to another chromosome. Down's syndrome is characterized by intellectual disability, delayed physical growth, and a characteristic face (small chin, slanted eyes, and flattened nasal bridge). Associated health problems include congenital heart disease, epilepsy, leukemia, and thyroid disease.

Cell division: mitosis & meiosis

THE PROCESS OF ONE CELL DIVIDING INTO TWO IS KNOWN AS MITOSIS. It involves the fragmentation of the nuclear membrane and the winding up of each of the chromosomes to form discrete chromatid pairs during prophase, which are then aligned along the equator of the cell during metaphase. Each one of a pair of chromatids is attached by spindle fibers to one of a pair of centrioles. During anaphase, the chromatid pairs separate and each member of the pair moves toward one of the centrioles. During telophase, the nuclear membranes reform and chromatids unwind. Formation of an intervening plasma membrane divides the two daughter cells. Between cell divisions the chromatids are copied to replenish the full complement of DNA.

Meiosis is the process whereby sex cells or gametes are produced. It is a more complicated process in that chromosome pairs separate in the first cell division (meiosis I). The chromatids of each chromosome then separate in the second meiotic division (meiosis II). This means that the final result is four daughter cells with only half the number of chromosomes—i.e., haploid cells—of the original diploid mother cell. These daughter cells or sex cells (sperm or ovum) will replenish their number of chromosomes by combining with another haploid sex cell during fertilization. Crossing over of chromosome pairs occurs during meiosis I (prophase I) to produce greater variety in gene combinations.

▶ **MITOSIS AND MEIOSIS**

Mitosis is the process of one cell dividing into two, whereas meiosis is the process of cell division used in the gonads (testes and ovaries) to produce sex cells. Meiosis I is the first part of the process in which crossing over of chromosome arms occurs to transfer genetic material between chromosome pairs. This increases genetic diversity in the offspring. Importantly, each member of a chromosome pair separates in meiosis I to go to different daughter cells. The telophase of meiosis I is followed by the prophase of meiosis II. Meiosis II is a bit like mitosis in that each chromosome separates into chromatids that move to the daughter cells. The end result of meiosis I and II is four haploid daughter cells with half the number of chromosomes of the original germ cell.

MITOSIS

Interphase—DNA
replication

Prophase—chromosomes
wind up and nuclear
membrane breaks up

Metaphase—chromosomes align
along the cell's equator

Anaphase—chromatids of each
chromosome separate and
move to opposite poles of cell

Telophase—two nuclei reform
and cytoplasm of daughter cells
begins to separate

Two daughter cells with
reformed nuclear membranes
and separate cell membranes

MEIOSIS I

Prophase I—chromosome
attachment and crossing over to
exchange genetic material
between chromosomes

Metaphase I—
chromosome pairs line
up along equator

Anaphase I—chromosome
pairs separate

Telophase I

Cytokinesis I

MEIOSIS II

Prophase II

Metaphase II—
chromosomes line up
along equator

Anaphase II—sister
chromatids separate

Telophase II

Cytokinesis II—four daughter cells with half
the number of chromosomes (i.e., haploid) have
been produced. Each daughter cell has only one
chromosome from each pair of homologous chromosomes.

Apoptosis

APOPTOSIS IS THE PROCESS OF
PROGRAMMED CELL DEATH. It often
occurs when cells are damaged in some
way, cannot complete the normal cell
cycle of growth and cell division, and
cannot repair themselves. Cells affected
in this way enter a programmed
pathway of cell suicide (intrinsic
pathway) that involves a type of
internal messenger known as Bcl-2 that
in turn causes the release of pro-
apoptotic proteins from mitochondria.

Intrinsic apoptosis is an important
part of normal development, e.g., the
formation of the spaces between the
digits in the embryo or the removal of
excess neurons from the developing
brain and spinal cord. Cells may also
be forced into apoptosis by an extrinsic
pathway. In that case, immune system
cells produce messengers that bind to
receptors (TNF, or Tumor Necrosis
Factor receptors) on the target cell
surface and activate enzymes called
caspases in the cell cytoplasm.

The end result either way is that
the DNA of the nucleus begins to
fragment and the cytoskeleton breaks
down. The dying cell begins to produce
blebs of cytoplasm containing
fragments of organelles and surface
ligands (binding points) that are
recognized by phagocytic cells
such as macrophages. This leads to
macrophages clearing away the debris
and recycling the materials from the
dead cell.

▶ **CELLULAR EVENTS IN
APOPTOSIS**

Cell death or apoptosis can be initiated
either by an internal mechanism that is
activated in response to cell injury or
by an externally signaled mechanism
due to interactions between cell
receptors and external ligands, i.e.,
signaling from outside the target
cell. Both mechanisms lead to
fragmentation of DNA, breakdown
of the cytoskeleton, and the formation
of membrane blebs. Cellular debris
is engulfed by phagocytes.

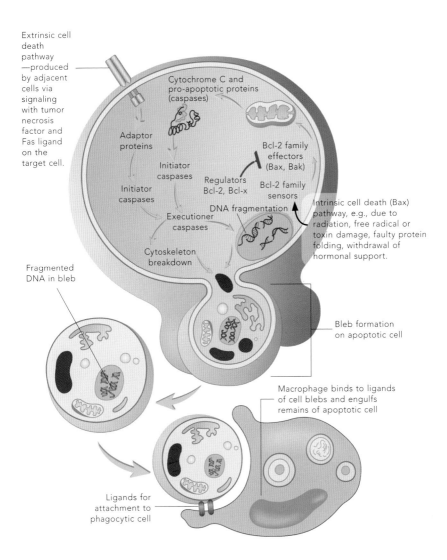

Extrinsic cell death pathway —produced by adjacent cells via signaling with tumor necrosis factor and Fas ligand on the target cell.

Cytochrome C and pro-apoptotic proteins (caspases)

Adaptor proteins

Initiator caspases

Initiator caspases

Regulators Bcl-2, Bcl-x

Bcl-2 family effectors (Bax, Bak)

Bcl-2 family sensors

DNA fragmentation

Executioner caspases

Cytoskeleton breakdown

Intrinsic cell death (Bax) pathway, e.g., due to radiation, free radical or toxin damage, faulty protein folding, withdrawal of hormonal support.

Fragmented DNA in bleb

Bleb formation on apoptotic cell

Macrophage binds to ligands of cell blebs and engulfs remains of apoptotic cell

Ligands for attachment to phagocytic cell

DNA replication

REPLICATION IS THE PROCESS OF DNA SYNTHESIS AND INVOLVES COPYING ONE DNA STRAND TO FORM THE MATCHING STRAND. Replication occurs in the period between cell divisions, during the synthesis or S phase of the cell cycle. Replication requires the unwinding of the chromatin of the chromosomes and the copying of the entire set of 3.2 billion base pairs in the human genome.

The key steps in replication are the unwinding and separation of DNA strands from the accompanying histone proteins in preparation for copying, the building of RNA primers on the strand to initiate copying, and the addition of nucleotides to the strand by the enzyme DNA polymerase. When DNA polymerase reaches the RNA primers, these are removed and replaced with DNA nucleotides. DNA synthesis proceeds in opposite directions along the two separated DNA strands.

The end result is two identical double helix strands of DNA. When DNA synthesis is complete, the histone proteins are recombined with the DNA strand, and the cell—with its full complement of DNA—then goes into the next stage of the cell cycle in preparation for the next cell division.

▶ **REPLICATION OF DNA DURING THE S PHASE OF THE CELL CYCLE**

DNA replication first requires the separation and unwinding of the two strands by the enzyme helicase so that each strand can be copied. The enzyme primase then builds RNA primers on the existing DNA strands so that the enzyme DNA polymerase has a starting point to begin copying each DNA strand. Each strand is then copied in opposite directions by DNA polymerase by the attachment of free nucleotides to the growing new DNA strands.

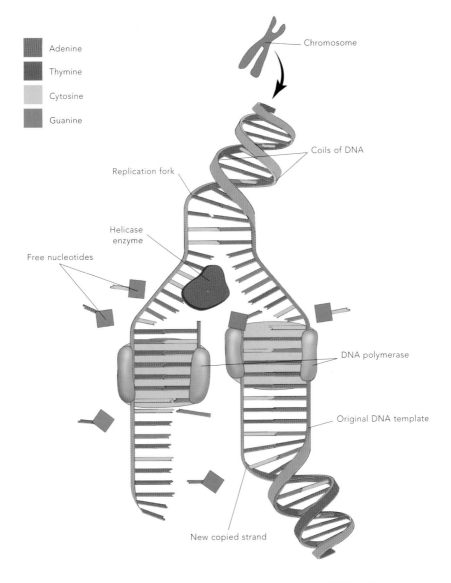

Adenine
Thymine
Cytosine
Guanine

Chromosome

Coils of DNA

Replication fork

Helicase enzyme

Free nucleotides

DNA polymerase

Original DNA template

New copied strand

Transcription & translation

TRANSCRIPTION IS THE PROCESS OF
CONVERTING THE CODED INFORMATION
IN THE DNA STRAND INTO AN RNA
MOLECULE. This can then leave the
nucleus and be used for protein
manufacture in the cytoplasm.
The process begins when transcription
factors bind to the promoter DNA
segment near the gene to be copied.
The enzyme RNA polymerase binds to
the promoter segment, and the segment
of DNA to be copied unwinds.

RNA polymerase then builds a
complementary strand of messenger
RNA, or mRNA (the mRNA
transcript), with nucleotides present in
the surrounding nucleus. Transcription
ends when the RNA polymerase
enzyme reaches the end of the gene
being copied and the mRNA transcript
is released.

Translation is the next step in the
process of making proteins. It involves
the use of mRNA to make a strand
of protein. The mRNA transcript
leaves the nucleus and moves to the
cytoplasm. Once there, the mRNA
strand is threaded into a ribosome,
which can be either free-floating or

attached to the rough endoplasmic
reticulum. Another type of RNA called
transfer RNA (tRNA) brings amino
acids from the cytoplasm to the
ribosome and mRNA strand. As each
tRNA molecule binds to the mRNA
strand, successive amino acids are
added to the protein strand. The
process is terminated when the
ribosome reaches the end of the
mRNA strand and the newly made
protein strand is released into the
surrounding cytosol.

▸ **TRANSCRIPTION OF DNA INTO
MESSENGER RNA**

Transcription requires the unraveling of
DNA and the construction of a strand of
messenger RNA (mRNA), using the DNA
of a particular gene as a template. After
transcription of the gene is complete,
the DNA strands are rejoined. The
mRNA is then transported to the
cytosol, where it is in turn used as a
template for protein construction in
the process of translation.

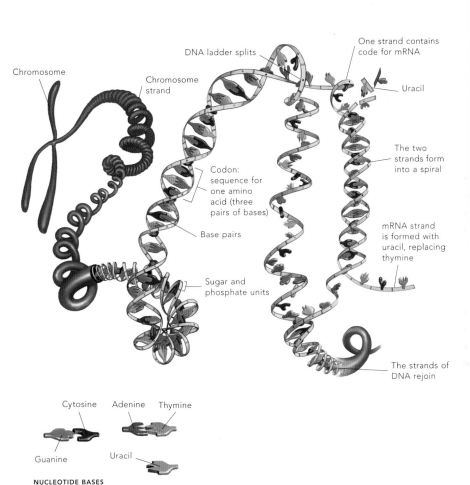

Chromosome

Chromosome strand

DNA ladder splits

One strand contains code for mRNA

Uracil

The two strands form into a spiral

mRNA strand is formed with uracil, replacing thymine

Codon: sequence for one amino acid (three pairs of bases)

Base pairs

Sugar and phosphate units

The strands of DNA rejoin

Cytosine   Adenine   Thymine

Guanine   Uracil

NUCLEOTIDE BASES

Energy production & ATP

ALL CELLS REQUIRE USABLE ENERGY TO RUN THEIR CELLULAR PROCESSES. The large molecules of proteins, fats, and complex carbohydrates that are ingested in food need to be converted to their constituents (amino acids, fatty acids, and glucose, respectively) before energy production can begin. Usable cellular energy then must come from exergonic (energy-producing) catabolic (breaking-down) reactions that convert amino acids, fatty acids, and simple carbohydrates to even smaller intermediate molecules like pyruvate or acetyl CoA, and ultimately to carbon dioxide and water.

All those catabolic reactions produce usable energy in the form of the adenosine triphosphate molecule (ATP) that can be transported throughout the cell to power energy-hungry reactions such as the synthesis of protein and structural fats. When the energy of ATP is used, the molecule is converted to adenosine diphosphate (ADP).

Glucose is converted to pyruvate in the cytoplasm by a process called glycolysis, but this produces only two ATP molecules per glucose molecule. The further metabolism of the pyruvate molecule occurs inside the mitochondria, where pyruvate is catabolized in the citric acid cycle to make acetyl-CoA and the reduced form of nicotinamide adenine dinucleotide (NAD)—NADH.

NADH then participates in the respiratory chain or oxidative phosphorylation reactions (see information on aerobic metabolism p. 48) in mitochondria to generate many more ATP molecules than protein catabolism, fatty acid oxidation, or glycolysis would alone. The energy-producing reactions that occur in mitochondria produce 34 ATP molecules per glucose molecule and release CO_2 and water.

▸ **SUMMARY OF ENERGY PRODUCTION FROM FOOD SOURCES**

Usable cellular energy in the form of ATP can be made from amino acids, fatty acids, or simple sugars like glucose. Anaerobic metabolism, which can occur in the cytosol of the cell without the input of oxygen, yields very little ATP per glucose molecule compared to aerobic metabolism, which occurs inside the mitochondria and requires oxygen.

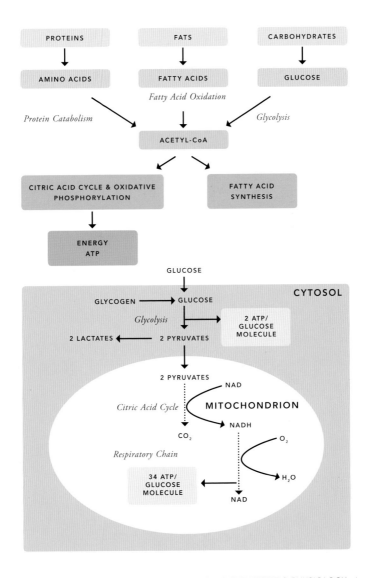

Aerobic metabolism

AEROBIC METABOLISM IS THE PART OF CELLULAR METABOLISM THAT REQUIRES OXYGEN. It occurs within mitochondria and consists of three stages: the transfer of electrons between electron carriers, the generation of a proton (H^+) concentration gradient, and the use of the proton gradient to drive the release of ATP. Electrons from NADH are transferred to molecules embedded in the inner mitochondrial wall, acting as a battery.

The electrical force built up by this electron transfer battery is used to pump protons against a concentration gradient. The protons are combined with oxygen to form water and also to drive the ATP synthase enzyme that generates ATP.

The complexity of this process— passing electrons between several different carriers—is necessary to generate useful energy without producing too much waste heat. Oxidative phosphorylation produces a far richer source of ATP molecules than any other metabolic process available to the body's cells and is why oxygen is such an important molecule for cellular survival.

The process of oxidative phosphorylation can be poisoned by cyanide, which binds to a key enzyme in the electron transfer chain, blocking its action and halting almost all ATP production.

▶ **OXIDATIVE PHOSPHORYLATION IN MITOCHONDRIAL INNER MEMBRANES**

Oxidative phosphorylation involves the transfer of electrons between 15 different carriers of the electron transport chain, the pumping of hydrogen ions or protons into the space between inner and outer mitochondrial membranes, and the combination of hydrogen ions or protons with oxygen to form water molecules.

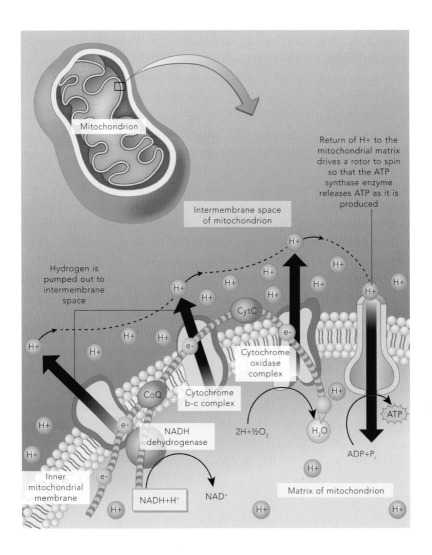

Mitochondrion

Return of H+ to the mitochondrial matrix drives a rotor to spin so that the ATP synthase enzyme releases ATP as it is produced

Intermembrane space of mitochondrion

Hydrogen is pumped out to intermembrane space

H+ H+ H+ H+ H+ H+ H+ H+ H+ H+ H+ H+

CytC

e-

Cytochrome oxidase complex

CoQ

e-

Cytochrome b-c complex

NADH dehydrogenase

$2H + \frac{1}{2}O_2$

H_2O

ATP

ADP+P$_i$

Matrix of mitochondrion

H+

Inner mitochondrial membrane

e-

e-

NADH+H$^+$

NAD$^+$

H+

H+

H+

H+

Cell membrane structure

CELL MEMBRANES CONTAIN TWO KEY COMPONENTS. The first of these is a double layer of phospholipids that is impervious to water. The bilayer consists of palisades of molecules (regular arrays like the palings on a picket fence) that have a "water-loving" (hydrophilic) head that faces toward the water-filled intracellular or extracellular fluid and chains of "water-hating" (hydrophobic) molecules that face each other within the membrane wall.

Embedded within the phospholipid bilayer are protein molecules that permit selective movement of specific chemicals across the lipid bilayer and serve cell-signaling functions.

Some types of membrane proteins span the entire thickness of the lipid bilayer (integral or transmembrane proteins), whereas others are found only on one side or another (peripheral proteins). Transmembrane proteins can serve as channels for small molecules like glucose and amino acids. Some can act as ion channels, allowing charged particles like sodium, potassium, or calcium ions to move across the bilayer.

Protein channels can be ligand-gated receptors, meaning they can be opened when a molecule attaches to a specific site on the channel protein molecule. Some membrane proteins act as enzymes, catalyzing specific chemical reactions. Others provide structural support by attaching to other structural proteins inside the cell or external to it. Yet other membrane proteins are linker proteins that bind one cell to another to form layered sheets.

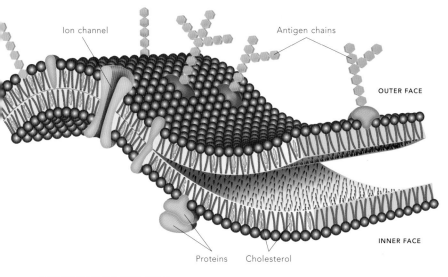

Ion channel

Antigen chains

OUTER FACE

INNER FACE

Proteins   Cholesterol

▲ STRUCTURE OF THE CELL MEMBRANE

All cells have a surrounding membrane that consists of a double layer of phospholipids with embedded proteins. The phospholipid bilayer has a hydrophobic core containing cholesterol and is impervious to water and most dissolved substances. Many molecules, e.g., ions, glucose, and amino acids, must cross the cell membrane through specialized channels, e.g., ion channels for sodium, potassium, chloride, or calcium. Proteins on the outer surface of the cell have antigen chains that allow for recognition by the immune system.

Movement across the cell membrane

THE MANNER IN WHICH DIFFERENT MOLECULES CAN MOVE ACROSS CELL MEMBRANES DEPENDS ON THE CHEMICAL CHARACTERISTICS OF THE MOLECULES. Cell membranes are made up of lipids (fats), so charged particles like ions or polar molecules such as glucose find it difficult to cross. Movement of chemicals can be passive, i.e., down a concentration gradient, or active, i.e., against a concentration gradient. Nonpolar molecules such as lipids (fats) and hydrocarbons, as well as gases like oxygen and carbon dioxide, can pass through the lipid bilayer by simple diffusion down a concentration gradient, i.e., from high concentration to low concentration.

Facilitated diffusion is the movement of charged particles like sodium ions and polar compounds such as glucose through protein channels or carrier proteins, but also down a concentration gradient. Active transport is an energy-expending process that moves ions or molecules across the cell membrane against a concentration gradient. Active transport requires the expenditure of energy in the form of ATP, so all active transport mechanisms are enzymes called ATPases.

The major primary active transport pump in the body is the sodium-potassium pump (also known as Na^+/ K^+ ATPase) and it directly uses ATP to power the pumping of these ions across the cell membrane against the natural concentration gradients. Secondary active pumps make indirect use of ATP. In other words, they use ATP to generate a concentration gradient of one substance, e.g., sodium ions that drive movement of another molecule, such as glucose.

▶ **MECHANISMS FOR MOVEMENT ACROSS THE CELL MEMBRANE**
Some lipid-soluble molecules and gases can cross the cell membrane by simple passive diffusion down a concentration gradient. Other polar compounds, e.g., glucose, can move passively across the cell membrane through special transport protein channels. Finally, some ions are moved across the membrane against concentration gradients by active transport mechanisms that require ATP.

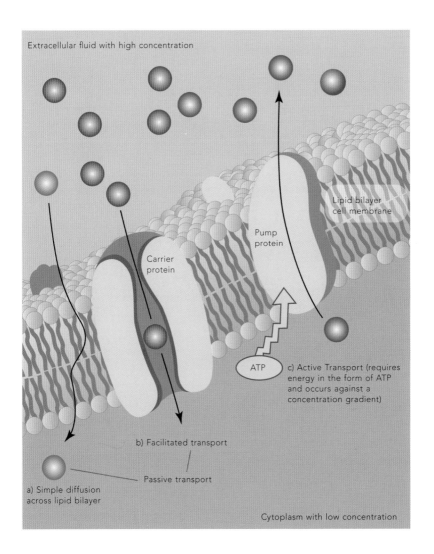

Extracellular fluid with high concentration

Lipid bilayer cell membrane

Pump protein

Carrier protein

ATP

c) Active Transport (requires energy in the form of ATP and occurs against a concentration gradient)

b) Facilitated transport

Passive transport

a) Simple diffusion across lipid bilayer

Cytoplasm with low concentration

The sodium-potassium pump

THE SODIUM-POTASSIUM PUMP IS
THE MOST COMMON TYPE OF ACTIVE
TRANSPORT PUMP IN THE HUMAN BODY.
This is because it performs the essential
task of maintaining the balance of
high sodium ion concentration
outside the cell and high potassium
ion concentration inside the cell. The
normal concentration of sodium is
10 times higher outside compared
to inside the cell, and the normal
concentration of potassium is 10
times higher inside compared to
outside the cell.

Without this concentration gradient,
contraction of heart and skeletal
muscle would be impossible and nerves
would be unable to send impulses,
because both of these processes depend
on allowing transient change in the
balance of charges across the cell
membrane using the ionic
concentration gradients.

The process of pumping sodium and
potassium ions requires expenditure
of energy in the form of adenosine
triphosphate (ATP). In fact, in some
cell populations, as much as 30% of
a cell's energy expenditure goes into

sodium-potassium pumping. The
sodium-potassium pump in the cell
membrane achieves this by moving
three sodium ions out of the cell for
every two potassium ions it moves
into the cell.

Despite the active pumping,
sodium and potassium constantly leak
across the cell membrane, so sodium-
potassium pumping is a continuous
process for the entire life of the cell.
If it fails, the cell swells and ruptures
as part of a process that is called
cellular necrosis.

▶ **THE SODIUM-POTASSIUM ACTIVE
TRANSPORT PUMP**

The sodium-potassium pump is an
energy-intensive mechanism for
maintaining the relative concentrations
of sodium and potassium on either side
of the cell membrane. It exchanges
three sodium ions moved out of the cell
for every two potassium ions entering
the cell, expending ATP in the process.

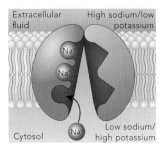

Step 1. Sodium binds to the sodium-potassium pump.

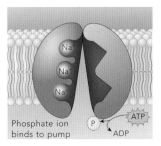

Step 2. Binding of sodium to pump stimulates conversion of ATP to ADP (diphosphate) and phosphate.

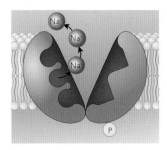

Step 3. Phosphorylation of the pump changes the protein shape, expelling sodium ions from the cell.

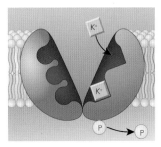

Step 4. Potassium from extracellular fluid binds to the pump, triggering the release of the phosphate.

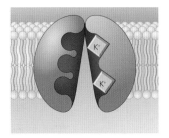

Step 5. Phosphate ion release changes protein shape back to the original.

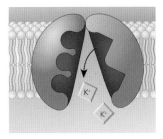

Step 6. Potassium is released to the cell interior and the cycle can repeat.

Membrane potential

THE MEMBRANE POTENTIAL OF A CELL IS THE ELECTRICAL POTENTIAL DIFFERENCE ACROSS THE CELL OR PLASMA MEMBRANE. Potential difference means that there is an electrical gradient across the membrane, and that the difference in charged particles would tend to drive positive ions into the cell if membrane channels were open. This means that the cell can be said to be polarized with respect to membrane potential, and that there is a potential for electrical work to be done.

The membrane potential can be measured by inserting an electrode through the plasma membrane and comparing the balance of charge inside and outside using a voltmeter. The normal membrane potential is usually about -70 to -90 mV (millivolts) for most excitable cells, e.g., nerve cells and muscle cells, at rest. By "at rest," we mean that the cell has not been electrically excited.

The membrane potential for all cells is dependent on the balance of charged particles on either side of the membrane and is achieved by more potassium ions being moved out of the cell than sodium ions are moved into the cell. In other words, the cell membrane is a little leakier for potassium than sodium ions. Note that the membrane potential of excitable cells, like nerve and muscle cells, can be changed for brief periods when ion channels open, as in the generation of an action potential (see pp. 72–73).

▶ **THE BALANCE OF CHARGES ACROSS THE CELL MEMBRANE**
There is an electrical potential difference across the cell membrane such that the inside of the cell is negative relative to the outside. This potential difference is maintained by the active movement of ions such that there are more positive charges outside than inside the cell.

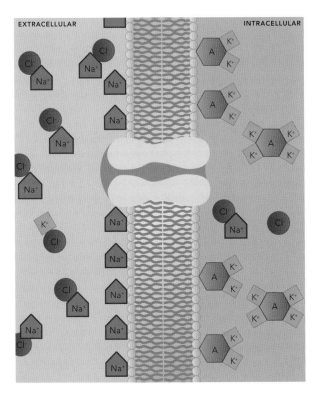

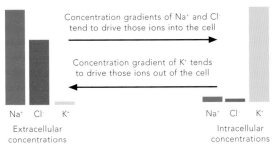

Concentration gradients of Na⁺ and Cl⁻ tend to drive those ions into the cell

Concentration gradient of K⁺ tends to drive those ions out of the cell

Na⁺ Cl⁻ K⁺
Extracellular
concentrations

Na⁺ Cl⁻ K⁺
Intracellular
concentrations

Cell surface & other receptors

ALL CELLS HAVE PROTEINS EMBEDDED IN THE LIPID BILAYER OF THEIR CELL MEMBRANE, AND MANY OF THESE PROTEINS SERVE IN COMMUNICATION BETWEEN CELLS. Some of these proteins are channel-linked receptors, meaning that an ion channel opens when a specific molecule binds to the receptor protein. Opening of the channel can change the membrane potential of the cell. An example would be the ion channels that open when a chemical neurotransmitter binds to a nerve cell at a chemical synapse.

Other cell surface receptors are linked with enzymes within the cell (enzyme-linked receptors). In this type, binding of a specific molecule to the receptor activates an enzyme that converts a substrate to a product. An example would be the tyrosine kinase receptors that are activated by growth factors, cytokines, and some peptide hormones.

G protein–coupled receptors can be bound by G proteins inside the cell when a chemical messenger also binds to the outside of the cell. The activated G protein can then bind to other enzymes like adenylate cyclase to produce a cellular response. This sort of activation is seen when hormones like epinephrine, glucagon, and calcitonin bind to cells.

Some signal receptors are located within the cell and are used by lipid-soluble molecules like steroid hormones that can diffuse across the cell membrane.

▶ **TYPES OF CELL SURFACE RECEPTORS AND SIGNALING MECHANISMS**

A variety of signaling mechanisms at the cell surface produce changes in the cell. These include channel-linked receptors that allow movement of ions in response to binding of chemicals to an external receptor, enzyme-linked receptors that activate an internal enzyme when a chemical attaches to the outside of the cell, and G protein–coupled receptors that bind and activate internal enzymes when a messenger chemical binds to the cell exterior.

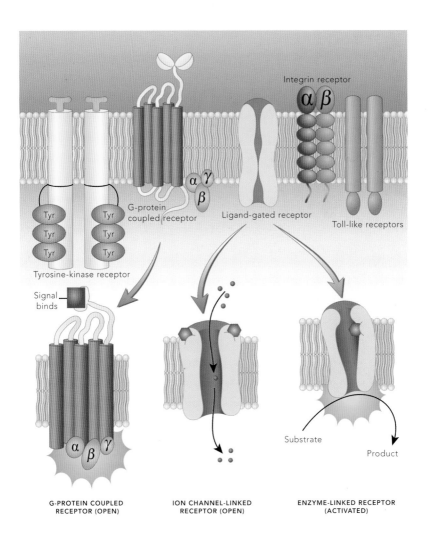

Integrin receptor

G-protein
coupled receptor

Ligand-gated receptor

Toll-like receptors

Tyr
Tyr
Tyr

Tyr
Tyr
Tyr

Tyrosine-kinase receptor

Signal
binds

α β γ

α β γ

Substrate

Product

**G-PROTEIN COUPLED
RECEPTOR (OPEN)**

**ION CHANNEL-LINKED
RECEPTOR (OPEN)**

**ENZYME-LINKED RECEPTOR
(ACTIVATED)**

Cell communication

CELLS COMMUNICATE WITH EACH OTHER IN MANY WAYS. Some communication is contact-dependent, meaning that a chemical signal bound to the membrane of one cell must be brought into contact with a specific receptor on the target cell. This type of communication is common in immune system responses when immune cells recognize damaged or malignant cells.

Some communication is at a local level and depends on local mediator chemical signals that diffuse only a short distance to target cells. This is called paracrine signaling and is seen within some cellular communities.

Synaptic signaling is used by nerve cells. In this type of communication, a chemical neurotransmitter is released from the axon terminal of one nerve cell into the gap of a chemical synapse to bind onto receptors on the dendrites, cell body, or axons of the target cell.

Endocrine signaling relies on diffusible chemical messengers (either peptide or steroid hormones) that are released from the cells of endocrine glands to be transported by the bloodstream or the fluid of body cavities to bind to receptors on target cells. When the chemical messengers, e.g., peptide hormones, bind to the target cell surface receptors, a signal-transduction pathway is initiated to activate specific cellular responses. Some cellular signals use messengers that can cross the cell membrane, e.g., steroid hormones, to bind to receptors within the cytoplasm.

▶ **MECHANISMS OF CELLULAR COMMUNICATION**

Cells must live together in communities, and this requires chemical interactions to occur between cells. Some of this communication is by direct contact-dependent interaction between cells, e.g., immune system communication. Synaptic contacts are essential for communication between nerve cells and between nerves and muscles. Endocrine signaling relies on diffusible agents that either bind to the cell exterior (peptide hormones) or cross the cell membrane to bind to receptors inside the cell (steroid hormones).

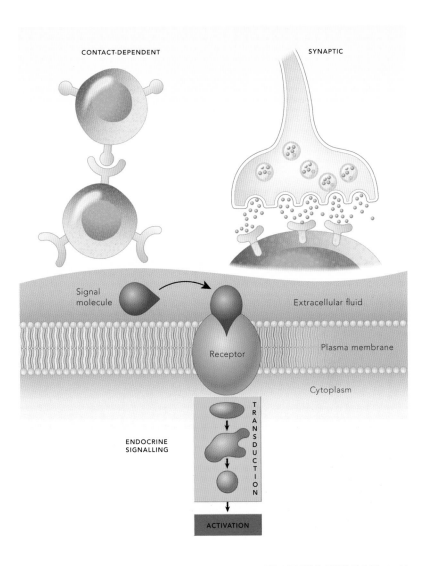

CONTACT-DEPENDENT

SYNAPTIC

Signal molecule

Extracellular fluid

Receptor

Plasma membrane

Cytoplasm

TRANSDUCTION

ENDOCRINE SIGNALLING

ACTIVATION

Functional organization of the nervous system

THE NERVOUS SYSTEM CAN BE DIVIDED INTO THE CENTRAL NERVOUS SYSTEM (brain and spinal cord) and the peripheral nervous system (peripheral nerves). About two-thirds of the brain is made up of the cerebral hemisphere, where much of the higher functions of the brain are located. By contrast, the medulla oblongata is concerned with more routine and automatic functions, such as respiratory and cardiovascular function.

The spinal cord has enlargements at the region of the base of the neck (cervical enlargement) for control of the upper limb and at the lumbosacral level (lumbosacral enlargement) for control of the lower limbs. The spinal cord ends at the level of the lower back, with nerve fibers below that point making up a bundle known as the *cauda equina* (Latin for "horse's tail").

Nerves of the peripheral nervous system are grouped into plexuses. The brachial plexus contains nerves to the upper limb, and the lumbosacral plexus supplies the lower limb. The main nerves of the upper limb are the axillary, radial, ulnar, and median, with digital branches to each of the fingers. Intercostal nerves supply the muscles of the chest wall. The three main nerves to the lower limb are the femoral, obturator, and sciatic nerves. The femoral nerves supply the muscles of the front of the thigh, the skin of the thigh, and the skin of the leg by the saphenous nerve. The obturator nerves supply the muscles and skin of the inner side of the thigh. The sciatic nerve supplies the muscles and skin of the back of the thigh and the leg and foot by its branches (tibial nerve and superficial and deep peroneal nerves, respectively).

▶ THE DISTRIBUTION OF PERIPHERAL NERVES

The peripheral nervous system consists of peripheral nerves to the trunk and limbs. The brachial plexus gives off nerves to the upper limb (axillary, radial, median, and ulnar). The lumbosacral plexus gives off nerves to the lower limbs (femoral, obturator, and sciatic). The sciatic nerve is the largest nerve in the body and provides tibial and peroneal branches. Intercostal nerves supply skin and muscle of the chest wall.

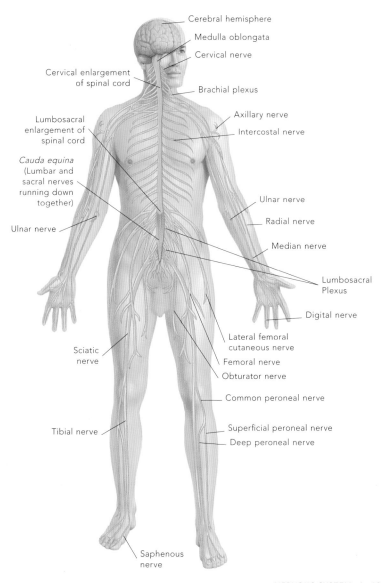

Cerebral hemisphere

Medulla oblongata

Cervical nerve

Cervical enlargement of spinal cord

Brachial plexus

Axillary nerve

Intercostal nerve

Lumbosacral enlargement of spinal cord

Cauda equina (Lumbar and sacral nerves running down together)

Ulnar nerve

Radial nerve

Median nerve

Ulnar nerve

Lumbosacral Plexus

Digital nerve

Lateral femoral cutaneous nerve

Sciatic nerve

Femoral nerve

Obturator nerve

Common peroneal nerve

Superficial peroneal nerve

Deep peroneal nerve

Tibial nerve

Saphenous nerve

Overview of brain structure & function

THE BRAIN CONTAINS A LARGE CEREBRAL CORTEX THAT CAN BE DIVIDED INTO FUNCTIONAL AREAS. Functional areas include the primary motor cortex, which is situated in the precentral gyrus and drives voluntary movement on the opposite side of the body, and the primary somatosensory cortex, which is found on the postcentral gyrus and is concerned with processing information about touch, pain, vibration, and joint position on the opposite side of the body.

The somatic sensory association cortex is located in the parietal lobe and constructs a three-dimensional model of the surrounding world. The motor or expressive speech area of Broca is located near the primary motor cortex in the left hemisphere in most people. The auditory cortex is located on the upper surface of the temporal lobe and is surrounded by the auditory association cortex and Wernicke's sensory speech area (usually in the left hemisphere). The visual cortex is located on the occipital lobe and is surrounded by the visual association cortex and a reading comprehension area. The diencephalon contains the thalamus, which is a way station for connections between the spinal cord or lower parts of the brain and the cerebral cortex. Cerebral peduncles contain descending connections from the cerebral hemispheres to the brainstem and spinal cord.

The brainstem (midbrain, pons, and medulla oblongata) is concerned with automatic function, such as respiration, cardiovascular control, and gastrointestinal function. It has major connections to the cerebellum through the superior, middle, and inferior cerebellar peduncles, as well as attachments for several cranial nerves.

▶ **FUNCTIONAL AREAS OF THE CEREBRAL CORTEX**
The bulk of the human brain is made up of the two cerebral hemispheres. The surfaces of these are divided into functional areas that are consistent in position from one person to the next.

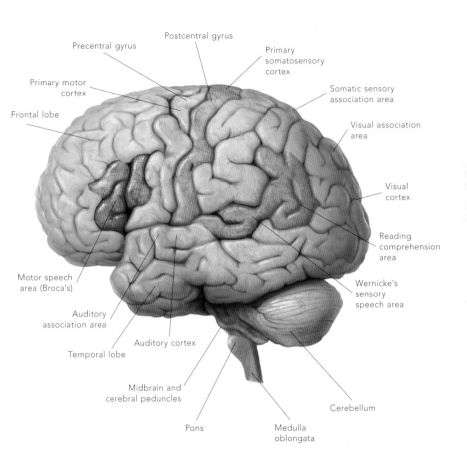

Precentral gyrus

Postcentral gyrus

Primary somatosensory cortex

Primary motor cortex

Somatic sensory association area

Frontal lobe

Visual association area

Visual cortex

Reading comprehension area

Motor speech area (Broca's)

Wernicke's sensory speech area

Auditory association area

Temporal lobe

Auditory cortex

Midbrain and cerebral peduncles

Cerebellum

Pons

Medulla oblongata

Fundamentals of brain development

THE CENTRAL NERVOUS SYSTEM DEVELOPS FROM A FLATTENED SHEET (the neural plate) that is folded at the end of the third week of development to produce a tubular structure. The head end of this tube expands to form three primary brain vesicles (forebrain, midbrain, and hindbrain vesicles), whereas the tail end of the tube will become the spinal cord. Cell division in the wall of this tubular brain produces all of the nerve cells of the adult brain and most of the glia or supporting cells. By eight weeks, the major parts of the brain are clearly visible.

Expansion of the forebrain vesicle to each side forms the pallium or cortex of the telencephalon, while the midline area of the forebrain vesicle becomes the diencephalon. The midbrain vesicle forms the mesencephalon with the distinctive corpora quadrigemina on its top. The hindbrain vesicle forms the metencephalon (future pons and cerebellum) and the myelencephalon (future medulla oblongata). The wedge-shaped space between the metencephalon and myelencephalon becomes the fourth ventricle, with the rhomboid fossa as its floor. The expansion of the telencephalon begins to form the lobes of the hemispheres by 21 weeks, when the lateral fissure is visible between frontal and temporal lobes.

Progressive expansion of the cerebral hemispheres leads to folding of the brain surface to make the gyri (bumps) and sulci (grooves). Major cortical features such as the central sulcus are visible by 30 weeks in utero, and all the gyri and sulci can be seen by the time the fetus is born (40 weeks).

▶ **EXTERNAL APPEARANCE OF THE EMBRYONIC AND FETAL BRAIN**
The brain develops from the front end of the neural tube, which expands and folds to produce the brain parts. Nerve cells are generated by cell division in the neural tube wall.

SPINA BIFIDA
The embryonic brain develops from a neural plate that folds to form a tube. If the neural tube at the tail end fails to close, a defect called spina bifida will result. Affected children may have paralysis of the lower limbs and be unable to control their bowel and urinary bladder.

8 WEEKS

Midbrain (mesencephalon)
Corpora quadrigemina
Future cerebellum
Rhomboid fossa
Forebrain (telencephalon)
Diencephalon
Medulla (myelencephalon)
Metencephalon

11 WEEKS

Pallium (telencephalon)
Mesencephalon
Cerebellum
Cerebral peduncle
Medulla (myelencephalon)
Pons (metencephalon)
Corpora quadrigemina

21 WEEKS

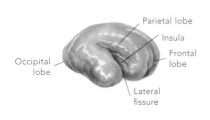

Parietal lobe
Insula
Frontal lobe
Occipital lobe
Lateral fissure

26 WEEKS

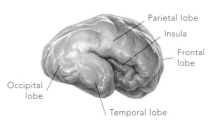

Parietal lobe
Insula
Frontal lobe
Occipital lobe
Temporal lobe

30 WEEKS

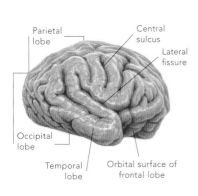

Parietal lobe
Central sulcus
Lateral fissure
Occipital lobe
Temporal lobe
Orbital surface of frontal lobe

40 WEEKS

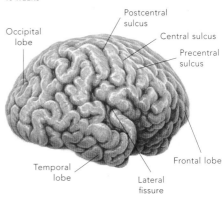

Occipital lobe
Postcentral sulcus
Central sulcus
Precentral sulcus
Temporal lobe
Lateral fissure
Frontal lobe

Overview of neuron structure & function

THE FUNDAMENTAL CELL TYPE IN THE NERVOUS SYSTEM IS THE NERVE CELL, OR NEURON. There are approximately 80 billion nerve cells in the adult human brain, most of which are in the cortex of the cerebral hemispheres and cerebellum. Neurons are concerned with transmitting and processing information in the form of electrical signals known as action potentials.

Most neurons have a polar structure, meaning that information enters the cell through synaptic contacts at one end (the dendrites or dendritic tree) and leaves via the other end (the axon). The summation of all the electrical inputs on the dendritic tree determines the overall electrical excitability of the neuron.

The outflow of information from the nerve cell body is through the axon, which is usually coated with a fatty myelin sheath to increase the speed of conduction of electrical signals to as high as 120 m/s (meters per second). The myelin sheath is interrupted at regular intervals by areas of naked axonal membrane called the nodes of Ranvier. When the nerve cell reaches a critical level of excitability (the

▶ **STRUCTURE OF THE TYPICAL NERVE CELL**

Nerve cells have a polar structure, meaning they have an input end (the dendrites) and an output end (the axon and axon terminals). Nerve cells contact other nerve cells through chemical synapses at the terminal boutons.

threshold), a wave of electrical change known as the action potential is initiated at the point where the axon begins from the cell body (the axon hillock). This impulse passes down the myelinated axon by saltatory propagation (literally leaping between nodes of Ranvier) to reach the axon terminal, where contact is made by chemical synapses with other nerve cells.

One nerve cell may make contact with as many as 50,000 other nerve cells, so the number of connections in the human brain is truly astronomical.

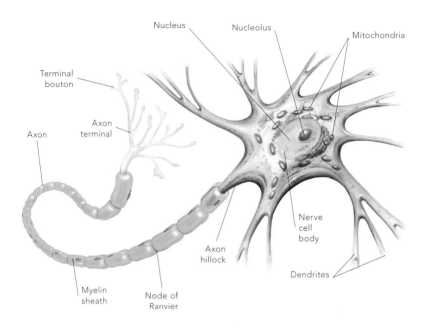

Nucleus

Nucleolus

Mitochondria

Terminal bouton

Axon terminal

Axon

Nerve cell body

Axon hillock

Dendrites

Myelin sheath

Node of Ranvier

EPILEPSY

Epilepsy is a broad group of conditions in which there are abnormal electrical discharges of nerve cell networks in the brain. Epilepsy may arise from the development of an abnormal region of electrically hyperactive brain tissue as a result of brain injury, tumor growth, or stroke, but it may also develop because of abnormal genes. The characteristic feature of epilepsy is the seizure, which classically consists of abnormal jerking or shaking of the body, e.g., grand mal seizure, but some epilepsy may show only as periods of apparent inattention called absence attacks, e.g., petit mal seizure.

Control of ion channels by membrane potential

VOLTAGE-GATED ION CHANNELS ARE CRITICALLY IMPORTANT IN THE GENERATION OF ACTION POTENTIALS, which are key events in the physiology of excitable cells such as nerve and muscle cells. Voltage-gated means the ion channel can be opened by changes in the membrane potential. There are two types of voltage-gated ion channels that are involved in action potentials: one for sodium ions and the other for potassium ions.

The voltage-gated sodium channel has three states (resting, activated, and inactivated) and two gates to regulate sodium ion flow (the inactivation gate and the activation gate). In the resting state, the activation gate is closed, but the inactivation gate is open. For ions to pass through the sodium channel, both the activation and inactivation gates must be open (activated state). Sodium ions then follow the concentration gradient with respect to those ions and flow into the cell. In the inactivated state—when the particular patch of axon is refractory to the generation of a second action potential—the activation gate is open, but the inactivation gate is closed, so sodium ions are still unable to flow into the cell. Each of these three states occurs at a particular stage in the generation of the action potential (see pp. 72–73).

Voltage-gated potassium channels have only two states (activated or rest) and one type of gate (activation gate open and activation gate closed). The opening of the potassium gate occurs in the later part of the action potential, after the sodium channels have activated and passed to the inactivated state.

▶ **FUNCTION OF VOLTAGE-GATED ION CHANNELS**

The membrane potential of excitable cells can be changed by the opening of ion channels. Opening of channels allows ions to move down their concentration gradient, making the inside of the cell transiently more positively charged, i.e., briefly changing the membrane potential to positive.

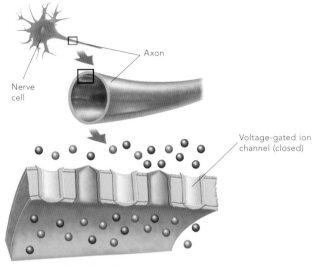

Axon

Nerve cell

Voltage-gated ion channel (closed)

INTERIOR OF AXON

Sodium ion channels closed
before or after action potential

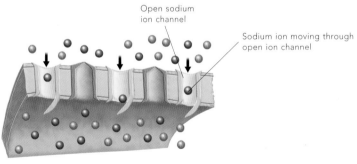

Open sodium
ion channel

Sodium ion moving through
open ion channel

INTERIOR OF AXON

Sodium ion channels open
during action potential

Action potentials: ionic basis of threshold, all-or-none response, & refractory period

AN ACTION POTENTIAL IS A RAPID CHANGE IN THE MEMBRANE POTENTIAL OF AN AXON OR MUSCLE FIBER THAT INVOLVES A DEPOLARIZATION, i.e., increase of the membrane potential, followed by a repolarization, i.e., the return to the resting state. Most importantly, action potentials spread down the axon (propagation) or along a muscle cell membrane in an all-or-none response, meaning that an action potential either happens completely or not at all—there is no intermediate state.

The events in an action potential start with a local increase in membrane potential (depolarization) at a trigger zone. This induces activation of voltage-gated sodium channels so that sodium ions enter the axon and the membrane depolarizes even more. This is a snowballing process: as the membrane potential depolarizes, more sodium channels open, depolarizing the membrane even more and opening more sodium channels. This proceeds until all sodium channels are activated and open and the membrane potential

has reached as much as +30 mV (millivolts). At this point the sodium channels inactivate, so sodium ions stop entering the axon, potassium channels begin to open, and potassium ions start to leave the axon.

The result is that repolarization, i.e., the return to the resting state, of the axonal membrane begins. As the membrane potential drops back toward the resting state, the sodium channels return to a resting state while the potassium channels remain open. This may tip the membrane potential below the resting level, i.e., as low as -80 mV, before the potassium channels inactivate and the membrane returns to a normal level of about -70 mV.

The axon is completely resistant to the stimulation of another action potential during the period when the sodium channels are either activated or inactivated (absolute refractory period) and relatively resistant during the period when sodium channels are at rest and potassium channels are activated (relative refractory period).

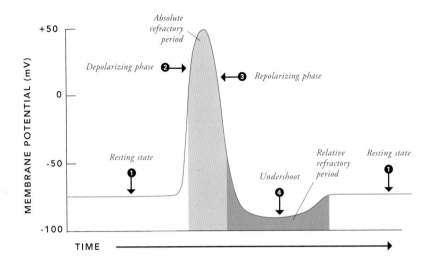

MEMBRANE POTENTIAL (mV)

+50

0

-50

-100

TIME

Absolute refractory period

Depolarizing phase ❷

Repolarizing phase ❸

Resting state ❶

Undershoot ❹

Relative refractory period

Resting state ❶

▲ THE COMPONENTS OF
THE ACTION POTENTIAL

The action potential is a transient change
in the membrane potential of the cell,
i.e., the membrane potential becomes
more positive. The depolarizing phase
involves the opening of sodium channels
in a snowballing process. When all
sodium channels are open, potassium
channels begin to open, commencing
the repolarizing phase. This causes the
membrane potential to return to negative
and even drop below the resting state
(undershoot). Eventually, the potassium
channels close, returning the membrane
potential to the resting state.

Action potentials: velocity & effects

IN THE CASE OF NERVE CELLS, the action potential normally spreads from a trigger zone on the cell body or the initial segment of the axon toward the axon terminals. In muscle fibers, an action potential spreads along the plasma membrane (sarcolemma) of the muscle fiber. The purpose of an action potential is to induce some change, whether that is the release of neurotransmitters from a chemical synapse (see pp. 76–77) or the initiation of muscle contraction (see pp. 174–175).

The rate at which the action potential propagates down the axon is called the conduction velocity. Conduction velocity depends on the diameter of the axon and the presence or absence of a myelin sheath. The wider the axon, the lower the internal resistance to current flow and the more rapid the conduction velocity. Nevertheless, there is a practical limit to axon diameter, so unmyelinated axons, e.g., C fiber pain and temperature axons, usually propagate action potentials at only 0.5 to 2 m/s (meters per second).

The presence of a myelin sheath makes saltatory (leaping) conduction along an axon possible because the insulating myelin sheath decreases the leakage of current from the axon interior to the outside, allowing depolarization to be confined to isolated patches of naked membrane (nodes of Ranvier) between the myelin segments. By restricting depolarization to occasional patches of axon membrane, the impulse can move much more rapidly down the axon. Large myelinated type A fibers are wide, have thick myelin sheaths, and conduct at up to 120 m/s.

▶ **PROPAGATION OF THE ACTION POTENTIAL**

The action potential is a moving wave of depolarization that spreads down an axon. Action potentials are initiated at the axon hillock, where depolarization of the membrane causes currents to flow to the next segment of axonal membrane. These currents trigger opening of voltage-gated sodium channels, causing depolarization there and spreading the zone of depolarization down the axon. Myelin coating allows the triggering currents to propagate further, so the depolarization only occurs at nodes of Ranvier and the action potential leaps down the axon from one node to the next.

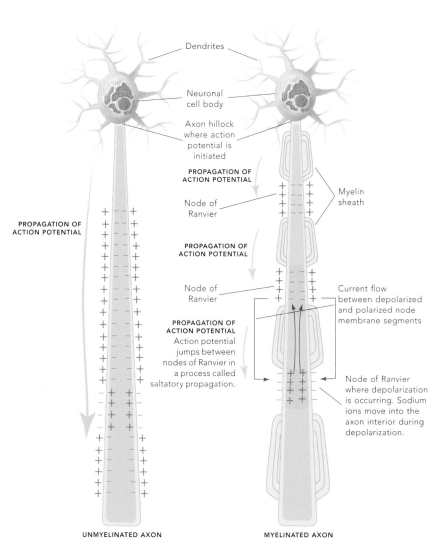

Dendrites

Neuronal cell body

Axon hillock where action potential is initiated

PROPAGATION OF ACTION POTENTIAL

Node of Ranvier

Myelin sheath

PROPAGATION OF ACTION POTENTIAL

PROPAGATION OF ACTION POTENTIAL

Node of Ranvier

Current flow between depolarized and polarized node membrane segments

PROPAGATION OF ACTION POTENTIAL
Action potential jumps between nodes of Ranvier in a process called saltatory propagation.

Node of Ranvier where depolarization is occurring. Sodium ions move into the axon interior during depolarization.

UNMYELINATED AXON

MYELINATED AXON

Synaptic structure & function

SYNAPSES ARE THE POINTS OF JUNCTION BETWEEN TWO NERVE CELLS. Some synapses simply consist of the two cell membranes in contact (electrical synapses) so that electrical activation of one cell spreads easily to the next. They transmit instantaneously and are bidirectional, i.e., transmission can go either way. Electrical synapses provide no opportunity for selectivity of response, so they are rarer than chemical synapses. Chemical synapses are also more efficient because there is no dissipation with distance as with an electrical signal.

A chemical synapse occurs where some part of the presynaptic cell (usually the axon terminal) discharges tiny packets or synaptic vesicles filled with neurotransmitters such as acetylcholine, dopamine, serotonin, glutamate, and aspartate into the synaptic cleft between the presynaptic and postsynaptic cells. The release of neurotransmitters is usually triggered when an action potential reaches the axon terminal and causes an influx of calcium ions into the axon terminal. The membrane of the postsynaptic cell may be part of a dendrite, cell body, or even another axon (axodendritic, axosomatic, and axoaxonic synapses, respectively). The postsynaptic membrane has specialized receptors on its surface that specifically bind the neurotransmitter chemical and induce some electrical change in the postsynaptic cell, usually by activating ligand-gated ion channels. When these channels are activated, either a local potential (postsynaptic potential) or an action potential is initiated in the postsynaptic cell, thereby transmitting information farther down the neural chain. After the postsynaptic cell has been activated, the neurotransmitter is either broken down, diffuses away, or is reabsorbed into the presynaptic cell.

PARKINSON'S DISEASE

Parkinson's disease is a chronic condition in which the dopaminergic neurons in the substantia nigra nucleus of the midbrain degenerate. Patients experience a pill-rolling tremor, muscular rigidity, and a masklike face because the forebrain is deprived of dopaminergic input.

▶ STRUCTURE AND FUNCTION OF A CHEMICAL SYNAPSE

A typical chemical synapse consists of a presynaptic axon terminal (sending nerve cell) contacting the dendrite of a receiving nerve cell. The presynaptic terminal releases packets of neurotransmitters into the synaptic cleft, where the molecules contact receptors in the postsynaptic membrane.

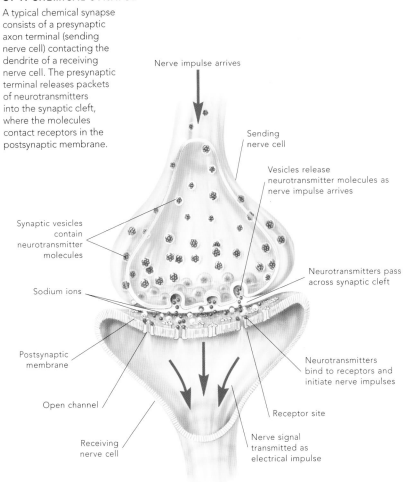

Nerve impulse arrives

Sending nerve cell

Vesicles release neurotransmitter molecules as nerve impulse arrives

Synaptic vesicles contain neurotransmitter molecules

Neurotransmitters pass across synaptic cleft

Sodium ions

Postsynaptic membrane

Neurotransmitters bind to receptors and initiate nerve impulses

Open channel

Receptor site

Receiving nerve cell

Nerve signal transmitted as electrical impulse

Synapses: excitatory & inhibitory

DEPENDING ON THE NEUROTRANSMITTER AND THE RECEPTOR, the binding of a neurotransmitter may either move the postsynaptic membrane potential closer to threshold (and hence initiation of an action potential) or farther away. If the discharge of neurotransmitter moves the postsynaptic membrane closer to threshold, it is said to have induced an excitatory postsynaptic potential (EPSP). If the discharge of neurotransmitter moves the postsynaptic membrane farther away from threshold, it is said to have induced an inhibitory postsynaptic potential (IPSP).

EPSPs are typically the result of the opening of sodium or calcium ion channels, with the entrance of positive ions into the postsynaptic cell and a rise in the normally negative membrane potential increasing the likelihood of an action potential. When EPSPs occur at the neuromuscular junction of the peripheral nervous system, they are called end-plate potentials. IPSPs usually result from the opening of potassium channels (with leakage of potassium ions out of the cell) or the

opening of chloride channels (with entry of negative chloride ions into the cell), with a drop in the normally negative membrane potential and a decrease in the likelihood of an action potential.

The sum of excitatory and inhibitory inputs on a given nerve cell (usually at different parts of its dendritic tree) determines the likelihood of the postsynaptic cell generating an action potential of its own. This is how nerve cells integrate or weigh up the many synaptic inputs they receive.

▶ **THE ACTIVATION OF THE GABA CHANNEL**

GABA (gamma amino-butyric acid) is a central nervous system neurotransmitter with inhibitory effects. Activation of a GABA channel occurs by binding of the neurotransmitter to the spaces between the alpha and beta subunits. Binding of GABA opens the ion channel to allow chloride ions to move through the center of the channel into the nerve cell. The drug benzodiazepine binds between the gamma and alpha subunits to enhance the effects of GABA.

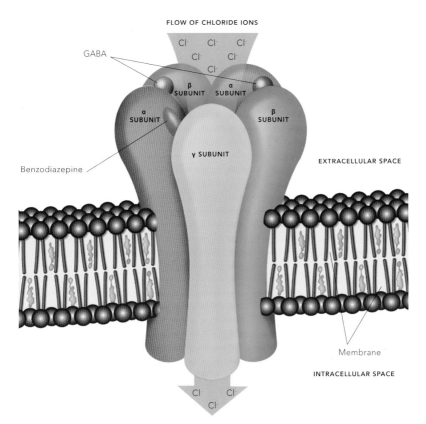

FLOW OF CHLORIDE IONS

Cl⁻ Cl⁻ Cl⁻
Cl⁻ Cl⁻
Cl⁻

GABA

β SUBUNIT

α SUBUNIT

α SUBUNIT

β SUBUNIT

Benzodiazepine

γ SUBUNIT

EXTRACELLULAR SPACE

Membrane

INTRACELLULAR SPACE

Cl⁻ Cl⁻
Cl⁻

Axoplasmic transport

TRANSPORT OF SUBSTANCES ALONG THE AXON IN BOTH DIRECTIONS IS ESSENTIAL FOR NERVE CELL FUNCTION. Proteins, mitochondria, and neurotransmitters made in the cell body of the nerve cell must be moved to the axon terminal, where they can provide structural support, generate energy, or be released into chemical synapses to act on other nerve cells. This is called anterograde transport.

It is also important for the nerve cell body to be able to sample the environments that are occupied by its axon terminals in a process called retrograde transport. This is essential for axon growth during development and for adult nerve cell survival. When a mature nerve cell is deprived of its connection with a target nerve cell or muscle, it may degenerate and die.

Axonal transport may be fast or slow. Slow axonal transport is only 1 to 3 mm per day and is used by cytoskeletal proteins. Fast axonal transport is at the rate of up to 400 mm per day and is used for transport of vesicles and membrane-bound organelles.

Fast transport proceeds along the neurotubular system of the axon and relies on motor proteins in the axoplasm that use ATP as an energy source. Retrograde axonal transport may be hijacked by some viruses, e.g., poliomyelitis, herpes simplex, and rabies, to infect nerve cells.

RABIES

Rabies is a viral infection of the brain caused by lyssa viruses. It is spread by infected saliva, when an animal bites or scratches a human. Most rabies is spread from dogs to humans, but bat bites can also transmit the disease. It is essential that an individual who has been bitten is given a vaccine immediately, but humans who have begun to show symptoms rarely survive.

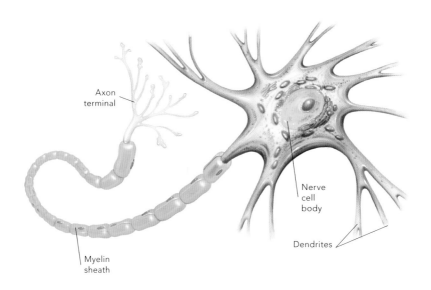

Axon terminal

Nerve cell body

Dendrites

Myelin sheath

▲ **TRANSPORT OF MATERIAL ALONG THE AXON**

Transport of materials along the axon in either direction is essential for nerve cell function and survival. Material manufactured in the cell body must be sent to the axon terminal for growth and the survival of the end of the axon and for release of neurotransmitters. Sampling of the environment of the axon terminal is essential for regulation of structured normal axonal growth and survival during development.

Sensory receptor cells & sensory transduction

SENSORY RECEPTOR CELLS ARE DIVERSE IN BOTH STRUCTURE AND FUNCTION. Sensation can be detected by nerve terminals themselves, e.g., pain and temperature, or by the interaction of a receptor cell with a nerve terminal, e.g., hearing and vestibular function.

The receptors of the retina (photoreceptors) are rods and cones, which both feature outer segments with laminated disk structures that contain light-sensitive pigment/protein complexes. Photoreceptors are tonically active, i.e., they are continuously active, producing an inhibitory neurotransmitter that inhibits the activity of bipolar nerve cells. When a photoreceptor is struck by a photon, this causes a drop in neurotransmitter release and removes the inhibition of bipolar cells.

Receptors for hearing and balance have hair processes (stereocilia and kinocilia) that can be bent by movement of fluids of the cochlea and vestibular apparatus. This bending of hair processes causes a receptor potential in the receptor cell that in turn triggers the release of excitatory neurotransmitters onto nearby sensory axons.

Olfactory and gustatory receptors are activated by binding of specific chemicals (odorants and tastants, respectively) onto receptor membranes. This binding activates a G protein/adenylate cyclase transduction pathway to induce a receptor potential and release excitatory neurotransmitters. Somatosensory receptors in the skin include relatively slowly reacting naked nerve terminals for pain and temperature perception, and sensory complexes like the Meissner corpuscle that is adapted for rapid response to stimuli for discriminative touch with fine spatial perception.

▶ **DIFFERENT TYPES OF SENSORY RECEPTORS**

Receptor cells come in many different forms depending on their function. Some are naked nerve terminals, e.g., pain and temperature, but most are specialized cells that in turn activate nerve endings by calcium-stimulated release of neurotransmitters.

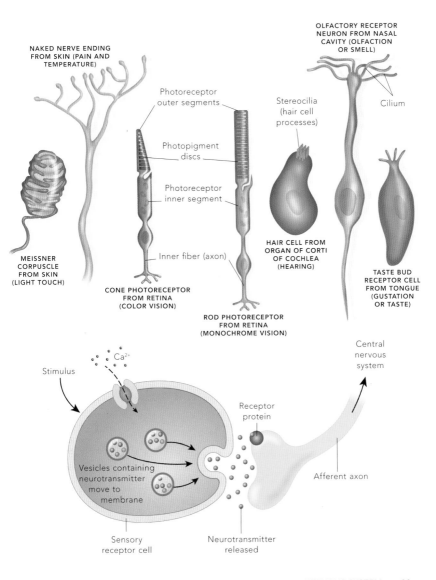

NAKED NERVE ENDING FROM SKIN (PAIN AND TEMPERATURE)

OLFACTORY RECEPTOR NEURON FROM NASAL CAVITY (OLFACTION OR SMELL)

Photoreceptor outer segments

Stereocilia (hair cell processes)

Cilium

Photopigment discs

Photoreceptor inner segment

Inner fiber (axon)

MEISSNER CORPUSCLE FROM SKIN (LIGHT TOUCH)

HAIR CELL FROM ORGAN OF CORTI OF COCHLEA (HEARING)

TASTE BUD RECEPTOR CELL FROM TONGUE (GUSTATION OR TASTE)

CONE PHOTORECEPTOR FROM RETINA (COLOR VISION)

ROD PHOTORECEPTOR FROM RETINA (MONOCHROME VISION)

Ca^{2+}

Stimulus

Central nervous system

Receptor protein

Vesicles containing neurotransmitter move to membrane

Afferent axon

Sensory receptor cell

Neurotransmitter released

Types of somatosensory receptors

THERE ARE MANY DIFFERENT TYPES
OF SENSORY RECEPTORS IN THE SKIN,
reflecting the different aspects of
somatosensation from the skin surface.
Naked or free nerve endings are the
structures in the skin that sense pain
or variation in temperature. They are
located within the epidermis itself so
that pain nerve endings also detect
separation of epidermal layers, e.g.,
in blister or vesicle formation as the
sensation of itch.

There are several different types
of touch receptors, each responding
to slightly different aspects of skin
contact. Merkel disks consist of nerve
endings surrounded by a capsule of
Merkel cells and are found where the
epidermis is thickest, e.g., skin ridges.
They are slowly adapting receptors,
meaning they give a sustained response
when a stimulus is applied for a long
period. They also provide fine spatial
resolution of touch information, i.e.,
discriminative touch, and are especially
common on the fingertips.

Krause end bulbs are embedded
in the dermis and also serve
discriminative touch, but with less

spatial resolution than the Merkel
disks. Ruffini endings are located in
the dermis and hypodermis and are
most sensitive to pressure, stretch,
and movement. They are part of
the proprioceptive (joint and limb
position) function of the skin.

Meissner corpuscles are rapidly
adapting, meaning that they are most
sensitive to changing tactile stimuli
and have a spatial resolution slightly
less than Merkel disks. Pacinian (or
lamellated) corpuscles are onion-
layered structures deep within the
dermis that respond optimally to
vibration. The layers act as a filter that
allows only high-frequency vibrations
to activate the receptor.

▶ **TYPES OF SOMATOSENSORY
RECEPTORS IN THE SKIN**
Sensation from the skin—known as
somatosensation—comes in many
forms. Pain and temperature are
detected by naked nerve endings.
Touch receptors respond to sustained
touch—e.g., Merkel discs—or vibration,
e.g., Pacinian or lamellated corpuscles.

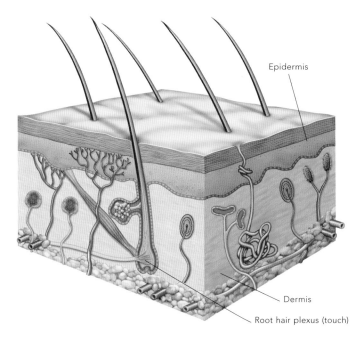

Epidermis

Dermis

Root hair plexus (touch)

Free nerve endings
(pain, heat, cold)

Krause end bulbs
(touch)

Meissner corpuscles
(touch)

Merkel disks
(touch)

Pacinian corpuscles
(vibration)

Ruffini endings
(pressure)

Spinal cord structure & function

THE SPINAL CORD IS THE PART OF THE CENTRAL NERVOUS SYSTEM THAT IS CONCERNED WITH THE INITIAL PROCESSING OF SENSORY INFORMATION FROM THE LIMBS, trunk, and internal organs and the motor control of the limbs, trunk, and viscera. It extends from the base of the skull to the level of the first lumbar vertebra in the middle back and is protected by the bony vertebral column. The spinal cord is divided into a central "H"-shaped region of gray matter, which contains nerve cell bodies and axon terminals, and the surrounding white matter columns, which contain ascending sensory (blue) and descending motor (red) axons that carry information up and down the spinal cord.

The gray matter is divided into the dorsal horn concerned with processing somatic sensory information, e.g., touch, pain, temperature, vibration, and joint position, and a ventral horn that contains motor neurons. The spinal cord has dorsal and ventral rootlets attached to it. Dorsal rootlets carry somatic sensory information into the spinal cord. The spinal or dorsal root ganglion contains the cell bodies of the sensory axons that run in the dorsal roots. Ventral rootlets contain axons of motor neurons destined for skeletal muscle for voluntary movement. The dorsal and ventral roots join to form a spinal nerve that immediately branches into a posterior ramus for muscles and skin of the back, and an anterior ramus for the limbs and front of the trunk.

Connective tissue sheaths of the spinal nerve include the epineurium, perineurium, and endoneurium. The spinal cord is surrounded by three layers of meninges. These are the pia mater on the spinal cord itself and the arachnoid mater and dura mater farther out from the spinal cord. The spinal cord has a rich vascular supply via anterior and posterior spinal arteries.

▶ **STRUCTURE OF THE SPINAL CORD**

This cross-sectional view of the spinal cord shows the outer coverings of the cord (pia mater, arachnoid mater, and dura mater), attached nerves, and the ascending (blue) and descending (red) pathways in the white matter.

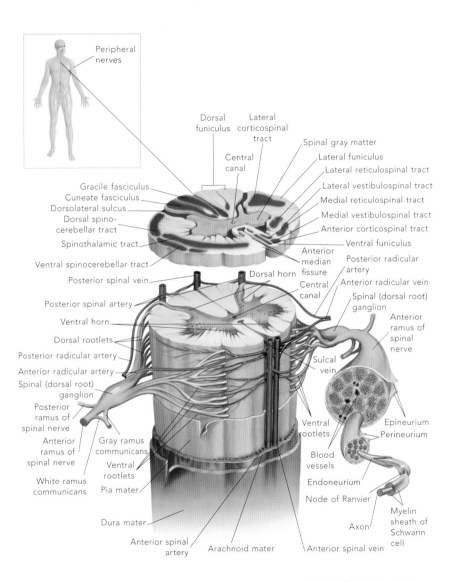

Peripheral nerves

Dorsal funiculus
Lateral corticospinal tract
Central canal
Spinal gray matter
Lateral funiculus
Lateral reticulospinal tract
Lateral vestibulospinal tract
Medial reticulospinal tract
Medial vestibulospinal tract
Anterior corticospinal tract
Ventral funiculus
Anterior median fissure
Posterior radicular artery

Gracile fasciculus
Cuneate fasciculus
Dorsolateral sulcus
Dorsal spino-cerebellar tract
Spinothalamic tract
Ventral spinocerebellar tract

Posterior spinal vein
Posterior spinal artery
Ventral horn
Dorsal rootlets
Posterior radicular artery
Anterior radicular artery
Spinal (dorsal root) ganglion
Posterior ramus of spinal nerve
Anterior ramus of spinal nerve
White ramus communicans
Gray ramus communicans
Ventral rootlets
Pia mater
Dura mater
Anterior spinal artery

Dorsal horn
Central canal
Anterior radicular vein
Spinal (dorsal root) ganglion
Anterior ramus of spinal nerve
Sulcal vein
Ventral rootlets
Blood vessels
Endoneurium
Node of Ranvier
Axon
Epineurium
Perineurium
Myelin sheath of Schwann cell

Arachnoid mater
Anterior spinal vein

Effects of spinal cord injury

SPINAL CORD INJURY IN HUMANS IS OFTEN AN IMPACT INJURY, e.g., a motor vehicle accident in which the vertebral column is crushed or shifted. Gunshot injury or stabbing with a knife can also cut through the spinal cord. Injury of the spinal cord can damage axons that descend or ascend the spinal cord and interrupt the flow of information along those axons.

This means that sensory information from parts of the body served by spinal segments below the injury will not reach the brain or conscious perception. It also means that motor commands from the cerebral cortex to the lower levels of the spinal cord will not get through.

When the spinal cord damage is at the level of the chest or abdomen, sensory information from the trunk and motor control of the lower limb, bowel, and urinary bladder will be lost. The motor paralysis in this case is known as paraplegia.

When the level of damage is higher (e.g., middle or lower neck), quadriplegia is the result, meaning that motor control of all four limbs is lost. There will also be no sensory information from any part of the body below the injury level, and voluntary control of the urinary bladder and bowel will be lost.

If the level of damage is even higher, e.g., at the top of the spinal cord, the motor neurons in the cervical cord that control the diaphragm will be separated from the brainstem's respiratory control centers, so activation of the muscular diaphragm will be lost and the affected individuals will be unable to breathe on their own.

▶ **THE PATTERN OF PARALYSIS FROM INJURY OF THE SPINAL CORD**
Damage to the spinal cord at the level of the first thoracic segment will result in paralysis of all parts of the body below that level, i.e., paralysis of the lower limbs—paraplegia—and loss of sensation of the trunk and lower limbs. A lesion higher up (middle cervical level) will cause paralysis of all four limbs (quadriplegia).

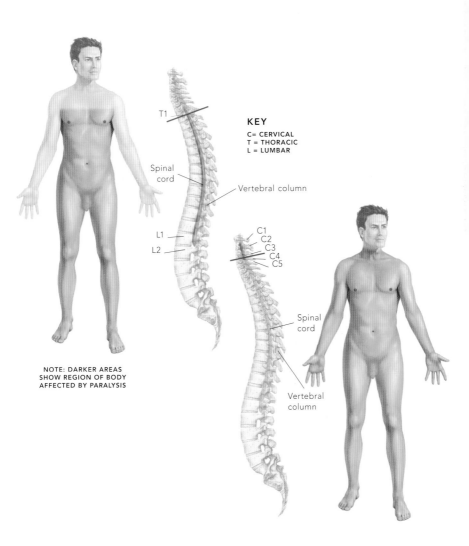

KEY
C= CERVICAL
T = THORACIC
L = LUMBAR

T1

Spinal cord

Vertebral column

L1
L2

C1
C2
C3
C4
C5

Spinal cord

Vertebral column

**NOTE: DARKER AREAS
SHOW REGION OF BODY
AFFECTED BY PARALYSIS**

Peripheral nervous system components

THE PERIPHERAL NERVOUS SYSTEM LIES OUTSIDE THE BRAIN AND SPINAL CORD AND CONSISTS OF GROUPS OF NERVE CELLS (ganglia) and nerve fibers (peripheral nerves). Ganglia may be either sensory in function, e.g., dorsal or sensory root ganglia, or autonomic, e.g., sympathetic trunk or celiac ganglion, controlling glands, and smooth muscle.

The peripheral nervous system is divided into sensory and motor divisions. Each of these divisions may be concerned with the exterior of the body and tissues of the limbs and body wall (somatic sensory or motor divisions), or with internal organs (visceral sensory or motor divisions).

Somatic sensory information is concerned with touch, pain, vibration, muscle stretch, and joint position sense (proprioception), whereas visceral sensory information is concerned with organ distension, e.g., the filling of the stomach or urinary bladder and traction on internal tissue folds (mesenteries).

The somatic motor division produces voluntary movement and consists of motor neurons in the ventral horn of the spinal cord or brainstem that directly drive voluntary skeletal muscle. The visceral motor division is part of the autonomic nervous system and is divided into sympathetic and parasympathetic divisions.

The sympathetic division is usually activated in emergency situations and expends energy rapidly to preserve the life and safety of the individual. The parasympathetic division is mainly concerned with routine vegetative function, e.g., the ingestion, digestion, and absorption of nutrients and the process of returning the internal environment to a resting state.

▶ MAJOR COMPONENTS OF THE NERVOUS SYSTEM

The nervous system is divided into the central nervous system (brain and spinal cord) and the peripheral nervous system (all nerves outside the brain and spinal cord). The peripheral nervous system provides motor and sensory function to the body wall and skin (somatic supply) and the internal organs (visceral supply).

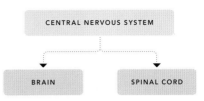

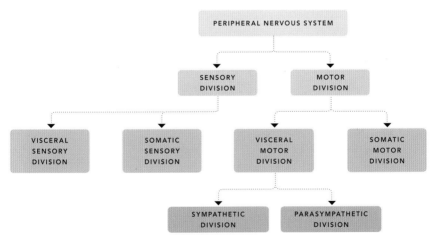

Cortical sensory regions

THE CEREBRAL CORTEX CONTAINS SEVERAL REGIONS CONCERNED WITH HIGHER PROCESSING OF SENSORY INFORMATION. The postcentral gyrus contains the primary somatosensory cortex, which is concerned with the perception of touch, conscious pain, vibration, and joint position. The primary somatosensory cortex is organized somatotopically, such that the face and upper limb are represented to the side and the trunk and lower limb are represented progressively closer to the midline.

The upper surface of the temporal lobe contains the primary auditory cortex concerned with hearing. This area is organized tonotopically, i.e., different areas respond best to different frequencies or pitches of sound. The primary visual cortex is at the back of the brain in the occipital lobe and is organized visuotopically, i.e., the different parts of the visual world are represented on different parts of the brain surface. The area of central vision has the largest representation and is located at the very back of the primary visual cortex.

Chemical senses such as smell and taste also have cortical areas, although they are small. The primary olfactory area is located on the inner surface of the temporal lobe, and taste is represented on the cortex inside the lateral fissure. In addition to these primary areas, there are also secondary or association cortex regions that deal with higher-order processing of sensory information, e.g., the color and visual form of objects and faces, or the position of objects in space.

▶ **THE MAJOR FUNCTIONAL AREAS OF THE CEREBRAL CORTEX**
The cerebral cortex is divided into functional areas concerned with motor activation (primary motor cortex) and sensory function (primary somatosensory cortex, visual cortex, and auditory cortex). Higher-order processing areas include the association cortex and the language areas (Broca's, Wernicke's, and reading comprehension or visuolexic area).

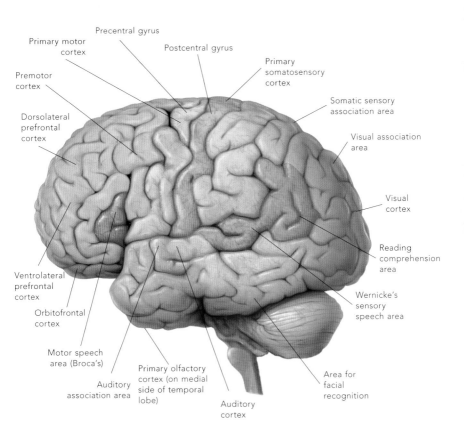

Primary motor cortex

Precentral gyrus

Postcentral gyrus

Primary somatosensory cortex

Premotor cortex

Somatic sensory association area

Dorsolateral prefrontal cortex

Visual association area

Visual cortex

Reading comprehension area

Wernicke's sensory speech area

Ventrolateral prefrontal cortex

Orbitofrontal cortex

Motor speech area (Broca's)

Primary olfactory cortex (on medial side of temporal lobe)

Area for facial recognition

Auditory association area

Auditory cortex

Somatosensation: pain & nociception

THE PERCEPTION OF PAINFUL STIMULI IS CALLED NOCICEPTION AND IS USUALLY PRODUCED BY TISSUE DAMAGE. Pain receptors in the skin are naked nerve terminals embedded in the epidermis. Pain, like the perception of temperature, is transmitted through peripheral nerves along the very smallest diameter axons with the least amount of myelin. This means that pain perception from some structures, e.g., viscera, is actually delayed by several seconds.

Primary pain axons have their cell bodies located in dorsal root ganglia, and their central axons terminate in the dorsal horn, where they contact the dendrites of neurons of the spinothalamic tract. These tract axons cross and ascend the spinal cord in the front part of the white matter before terminating in the ventral posterior nucleus of the thalamus, which has neurons that in turn project to the primary somatosensory cortex on the postcentral gyrus. The spinothalamic pathway is concerned with the discriminative aspects of pain sensation, e.g., location, intensity, and quality.

Some ascending axons terminate in the reticular formation of the brainstem on neurons that then project to diverse parts of the thalamus. These spinoreticulothalamic projections are concerned with the more unpleasant sensations associated with pain, such as unease and discomfort. Anti-inflammatory agents reduce pain by limiting the triggering of nociceptors in the peripheral tissues, whereas opioids act on nerve cells involved in the central perception of pain. The body's own opioids are the endorphins, which are used in a descending pathway from the periaqueductal gray of the midbrain to dampen the transmission of pain sensation in the dorsal horn.

NEUROPATHIC PAIN

Neuropathic pain is caused by damage to components of the pain sensory system, e.g., dorsal root ganglion cells or spinothalamic pathway, usually by the damage to a peripheral nerve. The condition is common (about 7% of the population) and causes abnormal sensations such as burning, stabbing, and itching.

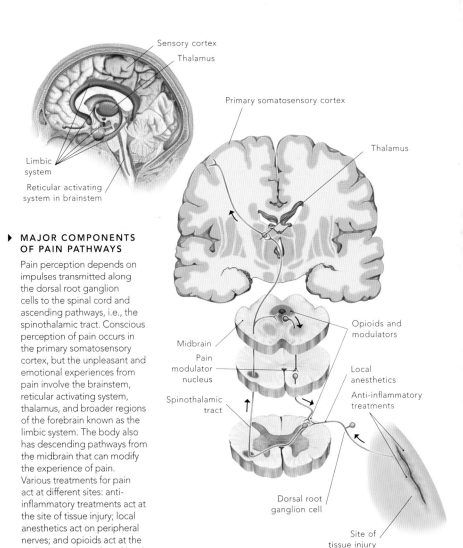

Sensory cortex
Thalamus

Limbic system

Reticular activating system in brainstem

Primary somatosensory cortex

Thalamus

Midbrain

Pain modulator nucleus

Spinothalamic tract

Opioids and modulators

Local anesthetics

Anti-inflammatory treatments

Dorsal root ganglion cell

Site of tissue injury

▶ MAJOR COMPONENTS OF PAIN PATHWAYS

Pain perception depends on impulses transmitted along the dorsal root ganglion cells to the spinal cord and ascending pathways, i.e., the spinothalamic tract. Conscious perception of pain occurs in the primary somatosensory cortex, but the unpleasant and emotional experiences from pain involve the brainstem, reticular activating system, thalamus, and broader regions of the forebrain known as the limbic system. The body also has descending pathways from the midbrain that can modify the experience of pain. Various treatments for pain act at different sites: anti-inflammatory treatments act at the site of tissue injury; local anesthetics act on peripheral nerves; and opioids act at the midbrain and spinal cord level.

Somatosensation: touch, pressure, vibration, & proprioception

TOUCH IS DETECTED BY A VARIETY OF RECEPTORS IN THE SKIN, e.g., Meissner's, Merkel's, and Pacinian receptors, and travels to the spinal cord (for trunk and limbs) or brainstem (for the face) by the axons of dorsal root or trigeminal ganglion cells, respectively.

Information for the conscious perception of discriminative touch, joint position, and vibration on the limbs and trunk is carried up the spinal cord in the axons of the dorsal columns (*fasciculus gracilis* and *fasciculus cuneatus*), which terminate in the brainstem nuclei (collections of nerve cell bodies) of the same name. The *nucleus gracilis* serves sensation from the lower parts of the body, whereas the *nucleus cuneatus* is for the upper parts. These two nuclei in turn send axons to the lateral part of the ventral posterior nucleus of the thalamus on the side of the body opposite to the sensation.

The medial or inner part of the ventral posterior nucleus of the thalamus also receives touch and vibration information, but from the face. The ventral posterior nucleus of the thalamus sends axons to the primary somatosensory cortex on the postcentral gyrus. The primary somatosensory cortex has a somatotopic organization, meaning that different parts of the body are represented in different regions of the cortex. The face and upper limb are represented on the side, and the trunk and lower limb are represented progressively closer to the midline of the brain. The face and hand have particularly large representations because of their behavioral importance.

▶ **COMPONENTS OF THE TOUCH, PRESSURE, AND VIBRATION PATHWAYS**

Fine or discriminative touch from the lower limb is carried by the *fasciculus gracilis* in the spinal cord white matter, whereas sensation from the upper limb is carried in the *fasciculus cuneatus*. Each of these pathways terminates in the respective nucleus in the medulla.

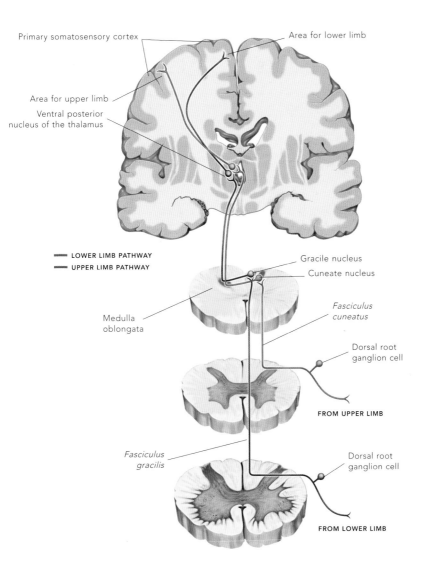

Primary somatosensory cortex

Area for lower limb

Area for upper limb

Ventral posterior nucleus of the thalamus

LOWER LIMB PATHWAY
UPPER LIMB PATHWAY

Gracile nucleus

Cuneate nucleus

Fasciculus cuneatus

Medulla oblongata

Dorsal root ganglion cell

FROM UPPER LIMB

Fasciculus gracilis

Dorsal root ganglion cell

FROM LOWER LIMB

Hearing: structure & function of the ear

THE EAR IS DIVIDED INTO THREE PARTS: an external part from the pinna to the tympanic membrane or ear drum; an air-filled middle part from the tympanic membrane to the vestibular and cochlear windows; and a fluid-filled inner ear, divided into the cochlea for auditory function and the vestibular apparatus for balance and the detection of linear and angular acceleration of the head.

The external ear directs sound from the external environment to cause vibration of the tympanic membrane. The series of auditory ossicles (malleus, incus, and stapes) serve as a mechanical device for amplifying movement of the tympanic membrane and transmitting that movement across the middle ear to the vestibular or oval window.

The middle ear is connected to the nasopharynx by the eustachian or auditory tube. That allows equalization of pressure between the middle ear and the external environment. Vibration of the oval window transmits pressure waves through the fluid of the inner ear, where they are detected by the sensory apparatus of the organ of Corti. Hair cells in that organ generate electrical signals that are transmitted to the brain through the cochlear nerve.

The vestibular apparatus contains the utricle and saccule for detection of head position and linear acceleration, and the semicircular ducts, which detect angular acceleration (rotation) of the head. The sensory parts of the vestibular apparatus are the maculae of the utricle and saccule and the ampullae of the semicircular canals. The inner and middle ear are embedded within the very dense petrous temporal bone that insulates the ear from extraneous sound.

▶ **COMPONENTS OF THE MIDDLE AND INNER EAR**

The ear is divided into three parts: external, middle, and internal. The middle ear is concerned with transmitting vibrations in the air to the fluid-filled interior of the inner ear, where sound is perceived.

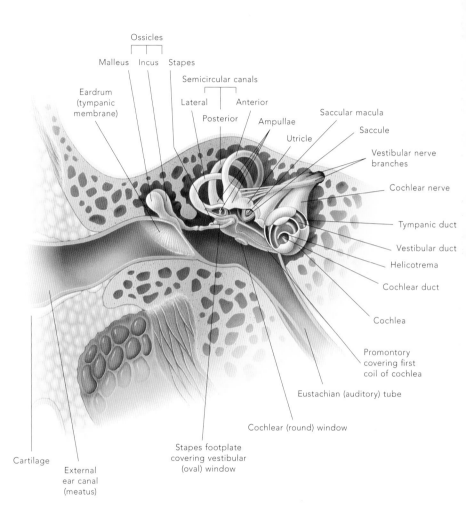

Ossicles

Malleus Incus Stapes

Eardrum
(tympanic
membrane)

Semicircular canals

Lateral Anterior

Posterior Ampullae

Saccular macula

Utricle Saccule

Vestibular nerve
branches

Cochlear nerve

Tympanic duct

Vestibular duct

Helicotrema

Cochlear duct

Cochlea

Promontory
covering first
coil of cochlea

Eustachian (auditory) tube

Cochlear (round) window

Stapes footplate
covering vestibular
(oval) window

Cartilage

External
ear canal
(meatus)

Hearing: physiology of the organ of Corti

THE YOUNG HUMAN EAR IS CAPABLE OF DETECTING SOUNDS WITH FREQUENCIES RANGING FROM 20 TO 20,000 HERTZ (cycles per second). Pressure waves generated at the vestibular window pass through the perilymph fluid of the inner ear along the spiral structure of the cochlea. Depending on their frequency (pitch) and wavelength, these sounds preferentially vibrate particular parts of the basilar membrane of the cochlea. High-frequency (short-wavelength) sounds travel only a short distance and cause the basilar membrane at the base of the cochlea (which is narrow and stiff) to vibrate; low-frequency sounds travel farther up the cochlea to induce vibration where the basilar membrane is wider and more flexible.

The organ of Corti has rows of inner and outer hair cells supported by rod and pillar cells. When the basilar membrane is vibrated, the stereocilia or hair process of the inner hair cells are bent relative to the overlying tectorial membrane, causing opening of potassium channels that depolarize the hair cells. This results in the release of excitatory neurotransmitters onto the axons of the cochlear ganglion cell, which initiates an action potential that travels to the brainstem.

Outer hair cells are predominantly amplifiers and can increase the movement of the basilar membrane as much as 100-fold to enhance the sensitivity of the ear to particular frequencies of interest. Outer hair cells are under the control of the brainstem.

▶ **STRUCTURE OF THE COCHLEA AND ORGAN OF CORTI**

Sound vibrations ascend the spiral of the cochlea (top right) through the fluid of the scala vestibuli, before crossing the cochlear duct to the scala tympani. The fluid vibrations cause movement of the basilar membrane and inner hair cells in the organ of Corti (lower left) leading to impulses in the cochlear nerve.

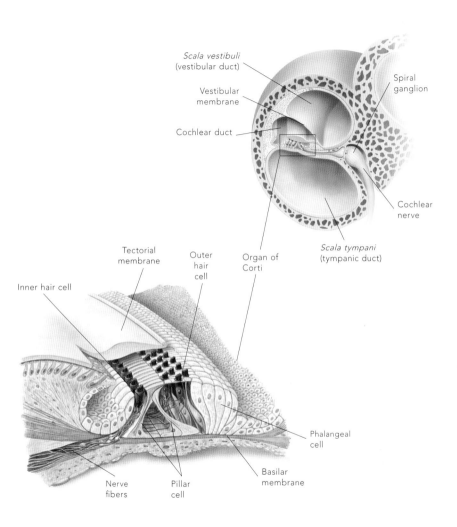

Scala vestibuli
(vestibular duct)

Vestibular
membrane

Cochlear duct

Spiral
ganglion

Cochlear
nerve

Scala tympani
(tympanic duct)

Tectorial
membrane

Outer
hair
cell

Organ of
Corti

Inner hair cell

Phalangeal
cell

Nerve
fibers

Pillar
cell

Basilar
membrane

Hearing: central processing of auditory information

THE INFORMATION ABOUT SOUND IS TRANSMITTED TO THE BRAINSTEM BY THE AXONS OF THE COCHLEAR DIVISION OF THE VESTIBULOCOCHLEAR NERVE. These axons terminate on nerve cells in the dorsal and ventral cochlear nuclei of the same side of the brain as the ear of origin. The cochlear nuclei in turn project to several auditory nuclei on both sides of the brainstem, e.g., superior olivary nuclei and nuclei of the lateral lemniscus, in an ascending pathway to the inferior colliculus of the midbrain. This means that information from one ear is passed to both sides of the brain very early in the pathway. The superior olivary nucleus is also the site of an outgoing pathway to the organ of Corti in the cochlea (olivocochlear bundle) that induces the amplifying actions of the outer hair cells.

The auditory pathway continues from the inferior colliculus to the medial geniculate nucleus of the thalamus. The medial geniculate nucleus nerve cells in turn project to the auditory cortex on the upper surface of the temporal lobe.

The primary auditory cortex is the site where sounds reach a conscious perception, and the region is tonotopically organized (see p. 92). This means that the lateral side of the auditory area responds best to low-frequency sound, and other areas to the front and back respond best to higher frequencies. Information streams from the primary auditory cortex to the sensory language area (Wernicke's) and areas of the temporal lobe concerned with musical appreciation.

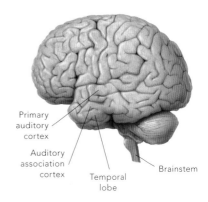

Primary auditory cortex

Auditory association cortex

Temporal lobe

Brainstem

▶ COMPONENTS OF THE AUDITORY PATHWAY

Auditory information from each ear passes up the brainstem on both sides, so that sound heard in either ear is perceived by both cerebral hemispheres. There are also many points of synapse between the many nerve cells in the auditory pathway, so a lot of auditory processing occurs before the information reaches the cerebral cortex. The auditory association cortex is concerned with the perception of different aspects of sound (e.g., music versus speech and the recognition of voices or birdsong).

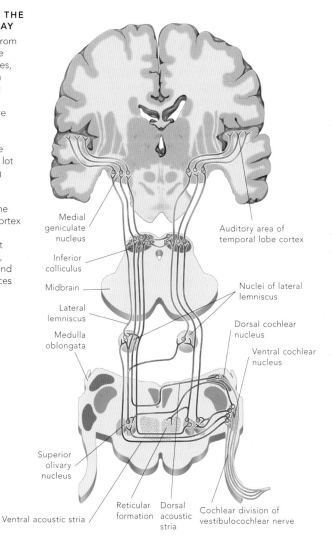

Medial geniculate nucleus

Auditory area of temporal lobe cortex

Inferior colliculus

Midbrain

Nuclei of lateral lemniscus

Lateral lemniscus

Dorsal cochlear nucleus

Medulla oblongata

Ventral cochlear nucleus

Superior olivary nucleus

Ventral acoustic stria

Reticular formation

Dorsal acoustic stria

Cochlear division of vestibulocochlear nerve

Vestibular function & balance

VESTIBULAR FUNCTION CAN BE DIVIDED INTO THREE GROUPS: the detection of static equilibrium of the head, including the sense of the downward direction; acceleration in a straight line (linear acceleration); and the sense of rotation of the head in space (angular acceleration). Static equilibrium and linear acceleration are detected by the maculae of the utricle and the saccule. The hair cells of these structures possess cilia (smaller stereocilia, plus a longer kinocilium) that are embedded in a gel containing crystals of calcium carbonate (otoliths) to increase the inertial mass. Movement of the head or the pull of gravity moves the gel and bends the cilia, initiating the depolarization of the hair cells and the generation of an action potential in the vestibular nerve axons.

Angular acceleration is detected by the semicircular ducts, which are three fluid-filled tubes (anterior, posterior, and lateral) orientated at right angles to each other. Each semicircular duct has a dilated segment known as the ampulla, which contains a crista ampullaris—a zone of hair cells with stereocilia embedded in an overlying cone of gel. Movement of fluid past the crista, which occurs when the head is rotated, bends the stereocilia, inducing depolarization of the hair cells and initiating action potentials in the vestibular nerve. Vestibular information is processed in the vestibular nuclei of the brainstem to influence eye movements and the balance of the body in space through the cerebellum.

MENIERE'S DISEASE

Meniere's disease is a condition of the inner ear that affects both hearing and balance. Affected individuals experience bouts of vertigo, tinnitus (ringing in the ear), and progressive hearing loss. It is due to the presence of excess fluid in the inner ear. Treatment may include restriction of salt in the diet and the taking of diuretics.

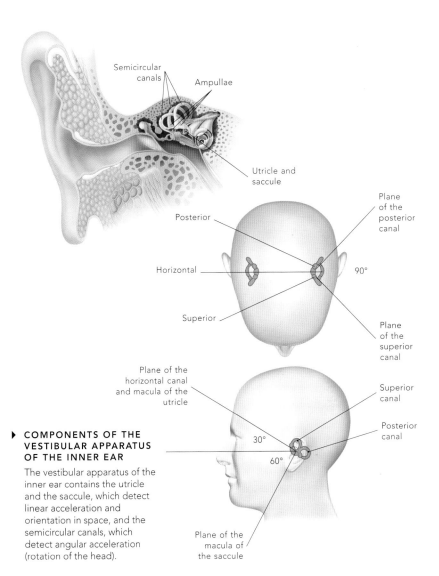

Semicircular canals

Ampullae

Utricle and saccule

Posterior

Plane of the posterior canal

Horizontal

90°

Superior

Plane of the superior canal

Plane of the horizontal canal and macula of the utricle

Superior canal

Posterior canal

30°

60°

Plane of the macula of the saccule

▶ **COMPONENTS OF THE VESTIBULAR APPARATUS OF THE INNER EAR**

The vestibular apparatus of the inner ear contains the utricle and the saccule, which detect linear acceleration and orientation in space, and the semicircular canals, which detect angular acceleration (rotation of the head).

Vision: structure & function of the eye

THE EYE CONSISTS OF A SERIES OF REFRACTIVE SURFACES AT ITS FRONT (cornea and lens) and a light-sensitive surface at its back (retina), as well as vascular support structures (the choroid and retinal vessels) and structural support (connective tissue of the sclera).

The external eye contains a conjunctival sac that keeps the eye moist with fluid produced by the lacrimal gland. The colored part of the eye is the iris, and its pigments protect the eye from solar radiation. The dark part of the external eye is the pupil, which is the adjustable-diameter optical entrance to the eye. The two optical elements of the eye are the cornea at the front and the lens behind it.

The muscles that adjust the size of the pupil are located in the iris (constrictor pupillae and dilator pupillae). The ciliary muscle adjusts the tension of the suspensory ligaments that pull on the equator of the lens. At rest, those ligaments are tight and the lens is flat; when the ciliary muscle contracts, the ligaments relax and the lens returns to a more globular shape. The retina is the light-sensitive neural structure of the eye

and is actually an extension of the brain. It is supported by vessels that run across its inner surface (retinal vessels) as well as by diffusion of nutrients from the very vascular choroid.

The fluid that fills the various chambers of the eye helps to maintain the eye's shape. These fluid-filled chambers are the anterior and posterior chambers around the iris, and the vitreous body behind the lens.

▸ **STRUCTURE OF THE EXTERNAL EYE AND EYEBALL**

Key components of the eye include a pair of optical elements at the front (the cornea and lens) and a light-sensitive surface at the rear (the retina). Information is transmitted to the brain along the optic nerve.

GLAUCOMA

Glaucoma is a group of conditions that damage the optic nerve. It may be due to raised pressure in the eyeball, but some affected individuals have normal or even low eye pressure. The condition is particularly serious because of the risk of vision loss.

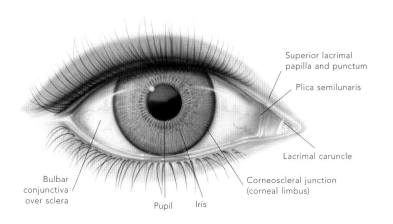

Superior lacrimal papilla and punctum

Plica semilunaris

Lacrimal caruncle

Corneoscleral junction (corneal limbus)

Bulbar conjunctiva over sclera

Pupil

Iris

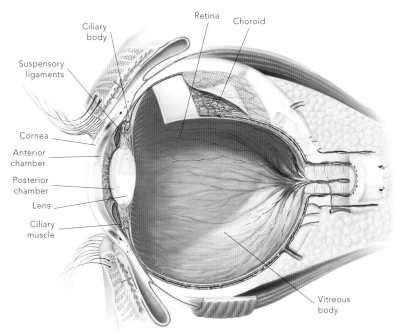

Ciliary body

Retina

Choroid

Suspensory ligaments

Cornea

Anterior chamber

Posterior chamber

Lens

Ciliary muscle

Vitreous body

Vision: optics of the eye

THE EYE FUNCTIONS AS A CAMERA, with optical elements at the front (cornea and lens) forming a completely inverted image on the light-sensitive retina at its back. The air/cornea interface is the major site for the bending of light rays, but the cornea cannot change shape and cannot focus. On the other hand, the lens is an adjustable optical element that can change its focal length. It becomes fatter (short focal length) when focusing on near objects and flatter (long focal length) when looking at distant objects. The tension of the suspensory ligaments of the lens can be adjusted by the muscle of the ciliary body, which contracts and releases the tension of the suspensory ligaments. This allows the lens to return to its naturally globular shape, when we focus on a near object.

The pupil of the eye is an adjustable aperture at the front of the eye. In low light levels, the pupil dilates by the action of the dilator pupillae muscle to allow as much light in as possible. In strong light levels or when focusing on near objects, the pupil constricts by action of the constrictor pupillae. The pupil also becomes smaller when focusing on near objects to minimize the effects of spherical aberration of the optical system when the lens takes on a globular shape.

The smooth muscles of the pupillary and ciliary muscles are under the control of the parasympathetic axons of the oculomotor nerve from the brainstem. The pupillary muscles take part in the pupillary light reflex, which automatically adjusts pupil size in response to light levels. Pupil dilation is under control of the sympathetic nervous system. The accommodation or focusing response is initiated in the cerebral cortex by voluntary control.

▼ IMAGE FORMATION AND TRANSMISSION TO THE CORTEX

The inverted image of the external world is formed on the retina by the cornea and lens. The retinal ganglion cells then transmit this information to the thalamus and visual cortex. The right hemisphere receives information about the left half of the visual world and the left hemisphere receives information about the right half of the visual world.

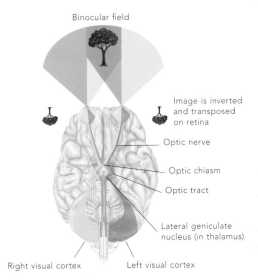

Binocular field

Image is inverted and transposed on retina

Optic nerve

Optic chiasm

Optic tract

Lateral geniculate nucleus (in thalamus)

Right visual cortex

Left visual cortex

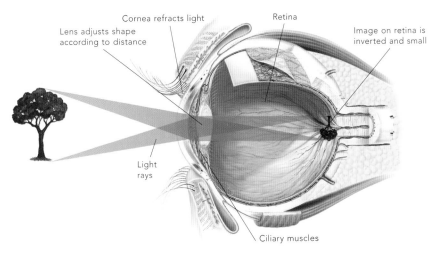

Cornea refracts light

Lens adjusts shape according to distance

Retina

Image on retina is inverted and small

Light rays

Ciliary muscles

Vision: phototransduction

THE PHOTONS IN THE LIGHT THAT REACHES THE RETINA MUST BE CONVERTED TO ELECTRICAL SIGNALS THAT CAN BE INTERPRETED BY THE BRAIN. This process is called phototransduction. Photons reach the photoreceptors of the retina by passing through all the layers of the retina.

The two types of photoreceptors are the rods and cones. Rods are the most numerous and are distributed across the retina. They are very sensitive and are capable of responding even in very low light levels. Cones contain pigments that allow the perception of color. They are concentrated in the fovea, which is the part of the retina that serves central vision, but are poorly responsive in low light levels. Both rods and cones have outer segments that house stacks of flattened light-sensitive disks formed from the cell membrane. In rods, the outer segment is cylindrical, whereas in cone photoreceptors, it is cone-shaped.

Each rod disk contains rhodopsin, which consists of the protein opsin and the pigment retinal. In rod photoreceptors at rest, the opsin and retinal are combined; sodium channels are opened in the outer segment; and the cell produces an inhibitory neurotransmitter called glutamate that continuously acts on the next cell in the chain, the bipolar cell. When a photon strikes the rod, the opsin and retinal separate, sodium channels close, and the photoreceptor reduces the release of glutamate. The result is that the bipolar cell is freed from inhibition and becomes more active, releasing an excitatory neurotransmitter that stimulates the next cell in the chain, the retinal ganglion cell, to produce an action potential that passes to the brain. So the resting state in the phototransduction mechanism is inhibition, and light releases the system from inhibition.

CELLS OF THE RETINA

Photoreceptor cells in the retina come in different shapes depending on their function (right). In a scanning electron micrograph (bottom), the outer segments of cone photoreceptors are purple and those of rod photoreceptors are gray.

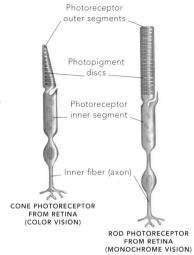

Photoreceptor outer segments

Photopigment discs

Photoreceptor inner segment

Inner fiber (axon)

CONE PHOTORECEPTOR FROM RETINA (COLOR VISION)

ROD PHOTORECEPTOR FROM RETINA (MONOCHROME VISION)

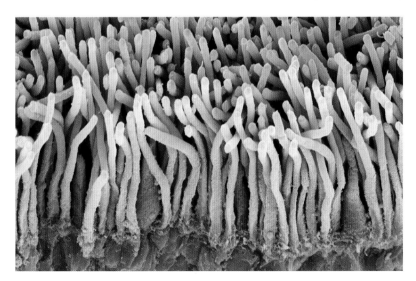

Vision: color vision

COLOR VISION DEPENDS ON THE SPECIAL PROPERTIES OF CONE PHOTORECEPTORS. There are three types of cones in the human retina: blue—maximally responsive to 420 nm (nanometers) wavelength light, green—maximally responsive to 530 nm light, and red—maximally responsive to 560 nm light. The disks in the outer segments of cones contain the pigment iodopsin and a protein called photopsin. Each of the cone types contains a different molecule of photopsin that absorbs different wavelengths. The color pigments respond to overlapping ranges of light wavelengths, so our perception of many types of color depends on the balance of input from the different cone channels.

When an individual lacks one or more functional genes for the cone pigments, color blindness is the result. The most common form of color blindness is due to the absence of genes for the red or green pigment, so the problem is called red-green color blindness. This is seen in as many as 10% of males but only 1% of females,

because the gene for color blindness is located on the X chromosome, of which males have only one but females have two. As long as one normal gene is present, color vision will be normal, so it is much more likely for males to have no functional pigment gene than females. Color blindness is tested by the Ishihara plates, which use a collection of different colored dots to produce an image of a number. Those with color blindness will see a different number from those with normal color vision.

▶ **THE ISHIHARA TEST OF COLOR VISION**

The Ishihara plates use a pattern of colored dots to convey an image that will be seen differently by those with normal vision compared to someone with color blindness. A person with normal color vision will read the number "75" in the plate to the right. No number will be visible to someone who has red-green color blindness.

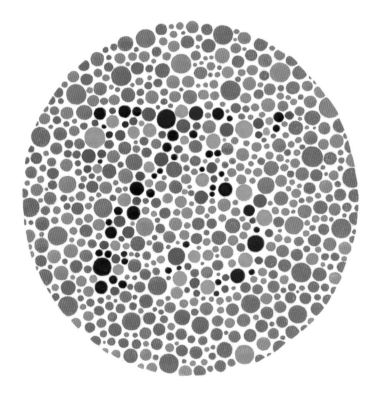

Vision: visual functions in the brain

VISUAL INFORMATION IS CARRIED BY RETINAL GANGLION CELL AXONS IN THE OPTIC NERVES TO THE BRAIN. Optic nerve axons from the nasal sides of each retina cross the midline in the optic chiasm so that information from the temporal or outer side of each visual field (remember that the image of the world is inverted and flipped left to right in the retina) is passed to the opposite side of the brain. Axons from the temporal retina stay on the same side of the brain. Processing of visual information continues in the lateral geniculate nucleus of the thalamus, before this information is passed on to the primary visual cortex in the occipital lobe for conscious awareness.

The transfer of visual information across the midline in the chiasm means that the left visual cortex receives information about the right half of the visual world and vice versa. The primary visual cortex is visuotopically organized, such that visual stimuli from the central part of the visual field are processed in the most posterior tip of the visual cortex. The primary visual cortex passes visual information on to other parts of the cerebral cortex in two streams: the dorsal stream to the parietal cortex concerned with the position of objects in space, and the ventral stream to the temporal lobe concerned with visual identification of objects (both inanimate objects and human faces). Information in the ventral stream also passes to the visuolexic (or reading) region of the left hemisphere at the junction of the temporal and parietal lobes.

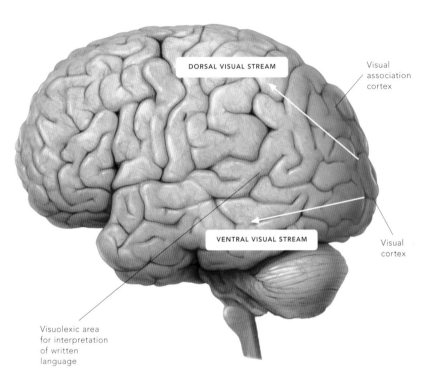

DORSAL VISUAL STREAM

Visual association cortex

Visual cortex

VENTRAL VISUAL STREAM

Visuolexic area for interpretation of written language

▲ **PROCESSING VISUAL INFORMATION IN THE CEREBRAL CORTEX**

The visual cortex is the first point in the cortex where visual information reaches conscious awareness, but other cortical areas process visual information further to map the positions of objects in visual space (dorsal visual stream) or recognize objects (ventral visual stream).

Sense of taste

THE SENSE OF TASTE, OR GUSTATION, DEPENDS ON THE BINDING OF CHEMICALS IN FOOD (tastants) with receptors in taste buds. Most taste buds are located on the upper or dorsal surface of the tongue, or on the palate between the oral and nasal cavities, but a few are present on the epiglottis behind the tongue. Taste buds are often found on protrusions of the tongue surface called papillae, including the 200 to 300 fungiform or mushroom-shaped papillae, the 15 foliate or tiger-striped papillae on the side of the tongue, and the 9 to 12 moat-shaped circumvallate papillae situated at the junction between the anterior and posterior parts of the tongue.

There are about 5,000 taste buds on the average tongue, and each taste bud contains three types of cells: the gustatory cell that contains the receptors for tastants; basal cells that differentiate into gustatory cells; and supporting cells that provide physical support for the gustatory cells. Sweet tastes are due to simple sugars like glucose and fructose. Sour is due to

the protons in food acids. Salty flavor is due to sodium and potassium ions. Bitter flavors are due to compounds in rancid or poisonous foods, while umami (savory) flavor is due to amino acids in food.

Information about taste is carried to the brainstem by the facial nerve for the front two-thirds of the surface of the tongue, the glossopharyngeal nerve for the back one-third of the tongue and the soft palate, and the vagus nerve for the epiglottis. The cortical area for taste is located in the lateral fissure.

▶ **TASTE BUDS AND TASTING REGIONS ON THE TONGUE**

Most taste sensation is perceived on the upper or dorsal surface of the tongue, where some sensations are felt more strongly in some regions compared to others. The taste buds are bundles of cells on the tongue surface, with specialized gustatory cells that activate nerve fibers to the brainstem.

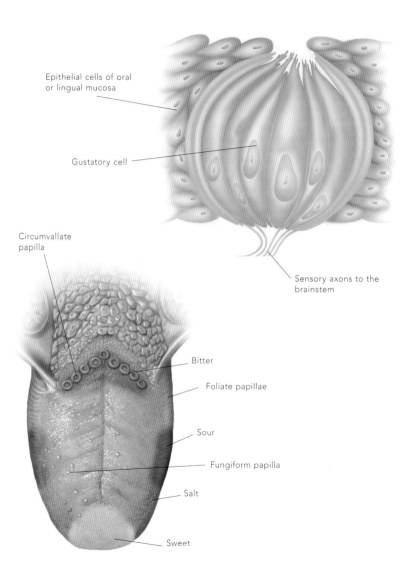

Epithelial cells of oral or lingual mucosa

Gustatory cell

Sensory axons to the brainstem

Circumvallate papilla

Bitter

Foliate papillae

Sour

Fungiform papilla

Salt

Sweet

Sense of smell

THE SENSE OF SMELL DEPENDS ON THE
BINDING OF CHEMICALS IN THE INHALED
AIR (odorants) with receptors on the cilia
of olfactory receptor neurons located in
the olfactory mucosa at the top of the
nasal cavity. The binding of odorants
with the receptors on olfactory neurons
is an example of a G protein–coupled
pathway that uses the enzyme adenylate
cyclase to signal the opening of ion
channels and trigger an action potential.

Olfactory receptor neurons send
olfactory nerve axons (cranial nerve I)
through the cribriform plate of the
ethmoid bone to penetrate the olfactory
bulb. The olfactory receptor neurons
are continuously exposed to the
external environment, so they must
be replenished by cell division in the
olfactory mucosa throughout life. The
olfactory mucosa also includes the ducts
of Bowman's glands that produce the
fluid essential for odorants to dissolve
so they can contact the processes of
olfactory receptor neurons.

Processing of olfactory information
occurs in the layers of the olfactory bulb
before information is transmitted back
to the brain by the olfactory tract axons
from mitral cells. The olfactory tract
ends in the primary olfactory cortex
on the inner side of the temporal lobe,
where olfactory input is able to influence
the function of the amygdala and the
hippocampal formation. Olfaction has
a direct influence on learning, memory,
and emotions by its connections with
these structures.

▶ COMPONENTS OF THE
OLFACTORY SYSTEM

The position of the olfactory areas
of the brain is shown above, and the
microscopic structure of the olfactory
mucosa and olfactory bulb is shown
below. Odor molecules (odorants)
contact the processes of olfactory nerve
cells in the olfactory mucosa, and
olfactory information is conveyed to
olfactory centers in the brain.

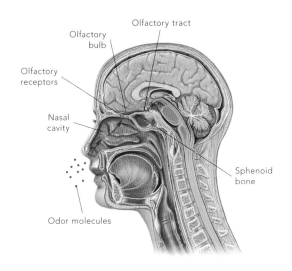

Olfactory tract

Olfactory bulb

Olfactory receptors

Nasal cavity

Sphenoid bone

Odor molecules

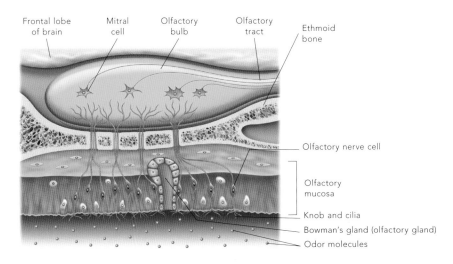

Frontal lobe of brain

Mitral cell

Olfactory bulb

Olfactory tract

Ethmoid bone

Olfactory nerve cell

Olfactory mucosa

Knob and cilia

Bowman's gland (olfactory gland)

Odor molecules

Spinal reflexes

SPINAL REFLEXES ARE STEREOTYPED RESPONSES TO SENSORY STIMULI IN THE SPINAL SENSORY SEGMENTS. Reflexes depend on a sensory neuron, some connections within the spinal cord gray matter, and one or more output motor neurons. The simplest reflex is the deep tendon stretch reflex, which causes contraction of a muscle when it is transiently stretched by a tap from a neurological hammer, e.g., the patellar ligament or knee jerk, and the Achilles tendon jerk. The circuit consists of only two neurons: a sensory axon carrying information from muscle stretch receptors and a motor neuron connecting to the skeletal muscle.

The next simplest reflex is the flexion withdrawal reflex that triggers withdrawal of a limb from hot or damaging agents. The withdrawal reflex involves at least two synaptic connections and may spread activation to several adjacent segments of the spinal cord. In this reflex, the painful stimulus triggers contraction of muscles to withdraw the limb, plus inhibition of antagonistic or opposing muscles. Other spinal segments away from those implicated in the initial stimulus may be involved because all muscles in the limb may be activated or inhibited in a vigorous response.

Superficial spinal reflexes include the plantar response that contracts the anti-gravity muscles of the calf and foot to push against the ground when the sole of the foot is stimulated. This reflex naturally assists in standing.

▶ **CIRCUITRY OF THE DEEP TENDON STRETCH REFLEX**

The deep tendon reflex circuit involves a stretch receptor in the muscle spindle, a sensory axon (purple axon), a single synapse in the spinal cord, and a motor output to the stretched muscle (red axon). There is also inhibition of the antagonistic muscle (blue axon).

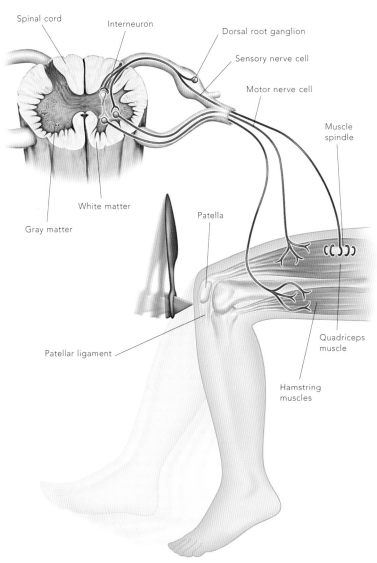

Spinal cord

Interneuron

Dorsal root ganglion

Sensory nerve cell

Motor nerve cell

Muscle spindle

White matter

Patella

Gray matter

Patellar ligament

Quadriceps muscle

Hamstring muscles

Overview of spinal motor pathways

THE BRAIN EXERTS CONTROL OVER SPINAL CORD MOTOR FUNCTION IN WAYS THAT CAN BE VERY PRECISE AND FINE, e.g., by the corticospinal tract from the primary motor cortex directly to the spinal cord gray matter, or by initiating preprogrammed motor routines that are stored in the reticular formation of the brainstem, e.g., by the corticoreticular and reticulospinal pathways. The former are particularly important in fine, independent, and skilled movements of the fingers, such as when playing the piano, whereas the latter are used for controlling posture and driving automated motor routines, such as swimming, walking, and running.

The reticulospinal pathways to the cervical and thoracic spinal cord also play a critically important role in regulating the respiratory rhythm. The rubrospinal pathway corrects motor patterns in the spinal cord in response to somatosensory feedback from the limbs and correction commands from the cerebellum, whereas the vestibulospinal pathway activates muscles essential for maintaining the balance of the trunk.

Each of these pathways descends along a different part of the spinal cord white matter. The corticospinal has two components with essentially similar function (lateral and anterior corticospinal tracts) that terminate directly on ventral horn motor neurons. The other descending pathways usually make contact with interneurons in the spinal cord gray matter, indirectly acting on motor neurons.

▶ **ORIGIN AND COURSE OF DESCENDING MOTOR PATHWAYS**

Motor pathways can originate in the cerebral cortex, i.e., the corticospinal tract for fine independent movement, with premotor and supplementary motor cortices directing particular motor routines (top illustration), or from the brainstem (rubrospinal, vestibulospinal, and reticulospinal tracts) for stereotyped and automated motor activity (bottom illustration).

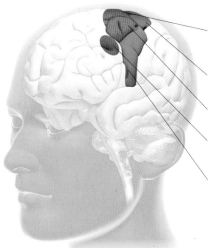

Supplementary motor area (on medial side of hemisphere): rhythmical movements, use of both arms together

Primary motor cortex: has direct control of motor nerve cells in spinal cord and brainstem

Upper premotor area: visual guidance of arm movements

Frontal eye field: moves eyes to opposite side

Lower premotor area: use of tools and interpreting other people's actions

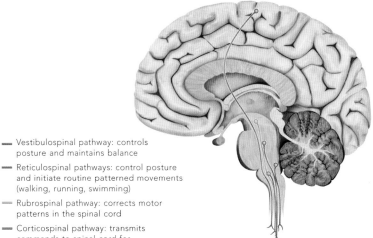

— Vestibulospinal pathway: controls posture and maintains balance

— Reticulospinal pathways: control posture and initiate routine patterned movements (walking, running, swimming)

— Rubrospinal pathway: corrects motor patterns in the spinal cord

— Corticospinal pathway: transmits commands to spinal cord for skilled movements

Motor control & coordination: basal ganglia & cerebellum

IN ADDITION TO THE DESCENDING PATHWAYS FROM THE CEREBRAL CORTEX AND BRAINSTEM NUCLEI TO THE SPINAL CORD, there are other brain centers that play important roles in motor control. The cerebellum lies posterior to the brainstem and is connected to the brainstem by three large peduncles or fiber bundles, feeding information into and out of the cerebellum. In general, the cerebellum is important in motor coordination, allowing the adjustment of muscle activation to maintain posture and balance and to produce smooth activation of muscles during complex movements.

The cerebellum is divided into three functional areas: a vestibulocerebellum that uses input from the vestibular apparatus to regulate head, neck, and trunk balance and eye movement; a spinocerebellum that uses somatosensory information from the spinal cord to regulate motor activation for posture and to compare actual limb movements with those required by descending cortical commands; and a pontocerebellum or cerebrocerebellum that participates in the planning and execution of fine skilled movements.

The basal ganglia circuits are also involved in motor control. They participate in feedback loops to the cerebral cortex that run from the motor cortex, through the basal ganglia, then the thalamus, and back to the cerebral cortex. This looped pathway helps ensure that movements are smooth. When damage occurs in the circuitry, individuals may exhibit tremors, dance-like jerks, or writhing movements. The motor function of the basal ganglia may be primarily to reinforce appropriate motor action and routines, while extinguishing those that are not appropriate.

THE FOREBRAIN SHOWING THE BASAL GANGLIA
HORIZONTAL SECTION

The deep parts of the forebrain contain the basal ganglia, large groups of nerve cells that engage in looped circuits involving the cerebral cortex, basal ganglia, and the thalamus. The basal ganglia include the caudate and putamen (forming the neostriatum) and the globus pallidus. The neostriatum receives dopaminergic axons from the substantia nigra.

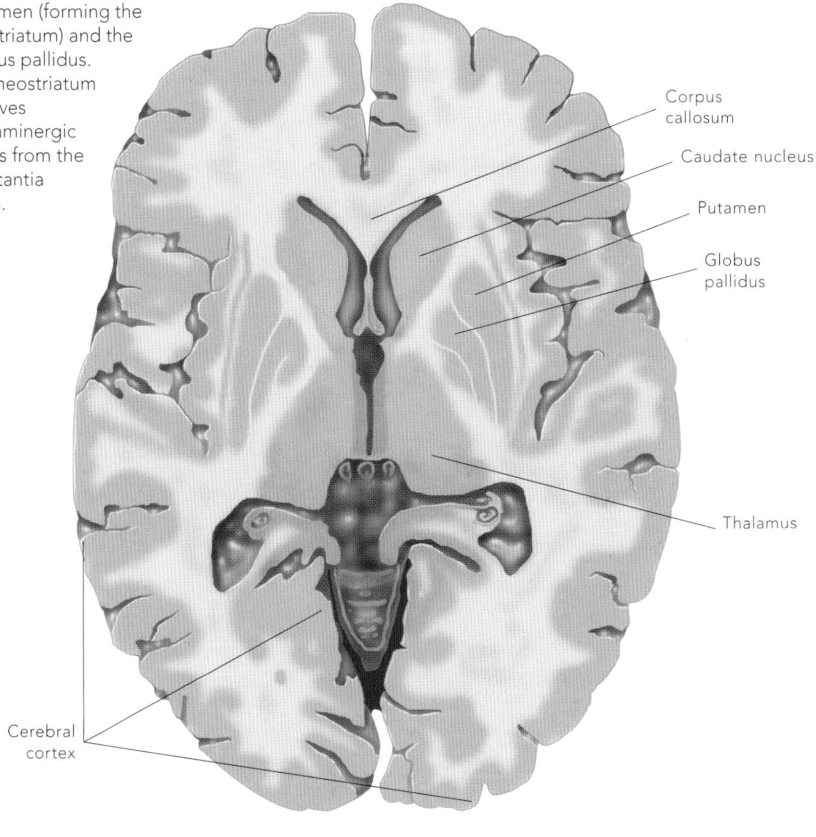

Corpus callosum

Caudate nucleus

Putamen

Globus pallidus

Thalamus

Cerebral cortex

Electroencephalograms

THE ELECTROENCEPHALOGRAM (EEG) is an investigation or medical test that analyzes the electrical activity of the surface of the brain. The individual wears an array of electrodes connected electrically to the scalp. The tracings of brain electrical activity are called brain waves and show characteristic changes during waking activity and during different stages of sleep, such as REM (rapid eye movement) and non-REM.

When the brain is awake and active and the eyes open, the characteristic brain waves (beta waves) are low amplitude and high frequency (13 to 30 cycles per second). When the person is awake but the eyes are closed, alpha waves are seen (8 to 12 cycles per second). Theta waves (higher amplitude and lower frequency than beta or alpha waves) are seen during the first three stages of non-REM sleep, while delta waves (very large amplitude and slow frequency—less than 4 cycles per second) are seen during stage four of non-REM sleep.

The EEG can be used to measure the rate of transmission of electrical activity along the visual pathways (visual evoked responses) from retina to visual cortex. A change in this rate may indicate loss of myelination of the optic pathways, as occurs in multiple sclerosis. The EEG is also an important test for diagnosing epilepsy, which is due to abnormal electrical foci within the cerebral cortex, and for detecting sleep disorders such as narcolepsy.

▶ **THE EEG IN THE DIFFERENT STAGES OF SLEEP**

The EEG undergoes characteristic changes as the individual passes from wakefulness to deep sleep. Essentially, the changes involve an increase in the amplitude and a lowering of the frequency of electrical activity as the subject progresses from stages one to four of non-REM sleep.

MULTIPLE SCLEROSIS

Multiple sclerosis is a debilitating condition in which the body mounts an immune attack against the myelin sheath that surrounds the axons of the central nervous system. When the myelin sheath is damaged, action potential conduction is much slower and cross-talk between axons may occur. Individuals with multiple sclerosis experience episodes of abnormal neurological function (e.g., blindness, pins and needles, loss of motor coordination) that may resolve, but the condition tends to progress over years. The EEG can be a useful tool in the diagnosis of multiple sclerosis in that it measures conduction velocity of the visual pathways, allowing the detection of abnormal patches of myelin (plaques) along those pathways.

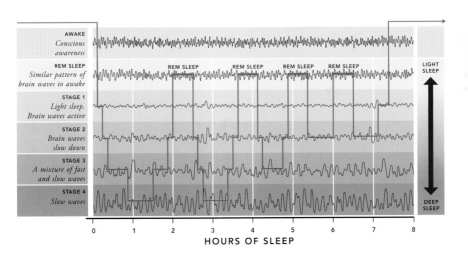

Sleep & wakefulness

SLEEP IS AN ACTIVE PROCESS, not an absence of function, and it is vitally important for health and optimal mental function. There are two types of sleep, which alternate through the night. Non-rapid eye movement (non-REM) sleep involves a series of stages during which the electrical waves on the electroencephalogram (EEG) become progressively slower and of higher amplitude. The eventual state of non-REM is stage four or slow-wave sleep, when the EEG shows large amplitude slow waves of less than 4 cycles per second. At this stage, muscle tone is at its lowest and parasympathetic activity is increased so that the heart rate and respiration are slow and steady. The underlying events of slow-wave sleep are cyclical electrical interactions between the cerebral cortex and thalamus.

About every 60 to 90 minutes, most people change to a period of desynchronized sleep activity. In this state, the EEG looks rather like the waking state, in that there is a great deal of low-amplitude but high-frequency activity. Despite the EEG similarities to the waking state, the individual is very difficult to rouse from sleep, and muscle tone and activity are completely lost. Breathing and heart rate become irregular, but the most striking feature is the presence of rapid eye movements (REM) that give this sleep state its name.

REM sleep is often accompanied by dreams, which involve activation of large areas of the sensory cerebral cortex, even though there is no external sensory input. By contrast, those parts of the cerebral cortex that are concerned with executive decision-making and social function are inhibited during dream sleep, accounting for the bizarre spatial and behavioral content of dreams.

▶ PATHWAYS FOR REGULATION OF SLEEP AND DAILY RHYTHMS

Sleep depends on ascending pathways from the brainstem and hypothalamus to the cerebral cortex (top illustration). The daily cycle of sleep and wakefulness is synchronized by exposure to light and by pathways through the nuclei of the hypothalamus, spinal cord, and sympathetic nervous system (see lower illustration). Secretion of melatonin from the pineal gland is highest during sleep.

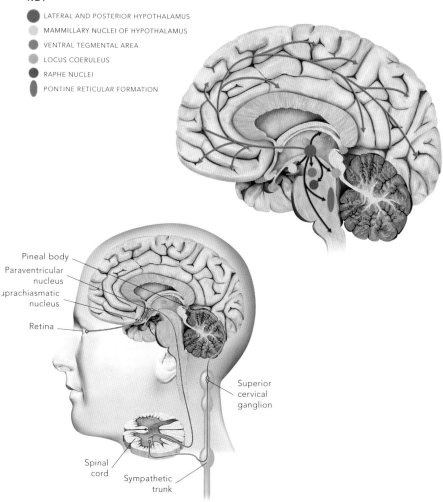

KEY

- LATERAL AND POSTERIOR HYPOTHALAMUS
- MAMMILLARY NUCLEI OF HYPOTHALAMUS
- VENTRAL TEGMENTAL AREA
- LOCUS COERULEUS
- RAPHE NUCLEI
- PONTINE RETICULAR FORMATION

Pineal body
Paraventricular nucleus
Suprachiasmatic nucleus
Retina
Superior cervical ganglion
Spinal cord
Sympathetic trunk

Emotions & the limbic system

THE LIMBIC SYSTEM IS THAT PART OF THE BRAIN THAT IS INTIMATELY CONCERNED WITH EMOTIONS AND MEMORY. The name "limbic" comes from the position of these structures around the edge (*limbus* in Latin) of the forebrain. The key structures of the limbic system include the amygdala and hippocampus in the temporal lobe, the mammillary bodies in the hypothalamus, the septal area in the basal forebrain, the anterior nucleus of the thalamus, and the cingulate cortex above the corpus callosum close to the midline of the brain. A looped circuit known as the Papez circuit connects many of these areas.

The amygdala is concerned with attaching emotional tags to memories and with making decisions on the basis of past emotional experiences, and the nearby hippocampus (hippocampal formation) is concerned with laying down new memories. The mammillary bodies play a key role in laying down memories, such that damage in chronic alcoholism leads to an inability to consolidate new memories. The septal nuclei and the nearby nucleus accumbens are involved in the pleasure and reward systems of the brain and play a critical role in addiction. The anterior nucleus of the thalamus and the cingulate gyrus are both important in the consolidation of new memories.

ALZHEIMER'S DISEASE

Alzheimer's disease is the most common form of a group of conditions called dementia, which means a progressive and irreversible decline in cognitive ability. Other types of dementia include Frontotemporal Dementia and Dementia with Lewy Bodies. In Alzheimer's disease, the brain shows abundant plaques of an insoluble substance called amyloid and neurofibrillary tangles inside nerve cells. Parts of the limbic system are damaged early in the course of Alzheimer's disease. In particular, the hippocampus is seriously affected, resulting in an inability to lay down new memories. Degeneration of the cerebral cortex leads to a relentlessly progressive loss of cognitive function.

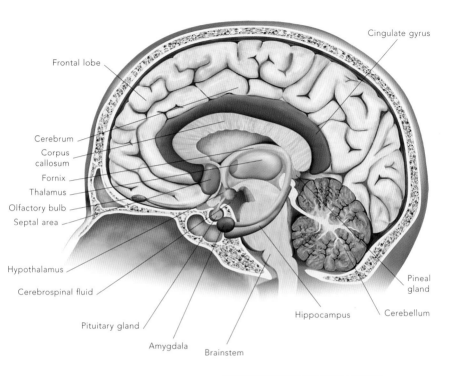

Cingulate gyrus

Frontal lobe

Cerebrum

Corpus callosum

Fornix

Thalamus

Olfactory bulb

Septal area

Hypothalamus

Cerebrospinal fluid

Pituitary gland

Amygdala

Brainstem

Hippocampus

Pineal gland

Cerebellum

▲ COMPONENTS OF THE LIMBIC SYSTEM

The limbic system is a loose collection of structures around the edge of the forebrain. Most are connected by a looped circuit called the Papez circuit.

Learning & memory

MEMORY COMES IN MANY FORMS. Working memory is the ability to remember a sequence of information or instructions while you perform a task. An example of this would be remembering a series of digits in a phone number while you dial. This relies on the dorsolateral prefrontal cortex at the front of the brain.

Longer-term memory can be divided into declarative and nondeclarative types. Declarative memory is that which can be consciously and verbally recalled, e.g., facts (semantic memory) or events (episodic memory). Nondeclarative memory is concerned with memories that do not involve factual statements. This could be memory for motor tasks (procedural memory) or for emotionally significant experiences and feelings (emotional memory).

Longer-term memories of the semantic, episodic, and emotional kind rely on the hippocampal formation in the temporal lobe. This part of the brain projects through a loop known as the Papez circuit to the hypothalamus, which in turn connects to the anterior nucleus of the thalamus and the cerebral cortex. Damage to the hippocampus and/or interruption of the Papez circuit can lead to profound anterograde amnesia, i.e., inability to remember information from the time of the damage forward.

The memories themselves are stored throughout the cerebral cortex. Emotional loading or charging of memories depends on a functional amygdala, which is located right in front of the hippocampal formation in the temporal lobe. Olfactory input also has a strong influence on emotionally tagged memories. Procedural memory depends on the basal ganglia.

▶ **PARTS OF THE BRAIN INVOLVED IN LEARNING AND MEMORY**

The hippocampus and its projections to the mammillary body of the hypothalamus are critically important in the laying down of new memories. Damage to either can result in profound problems in storing memories. The Papez circuit also involves the anterior nucleus of the thalamus and the cingulate gyrus.

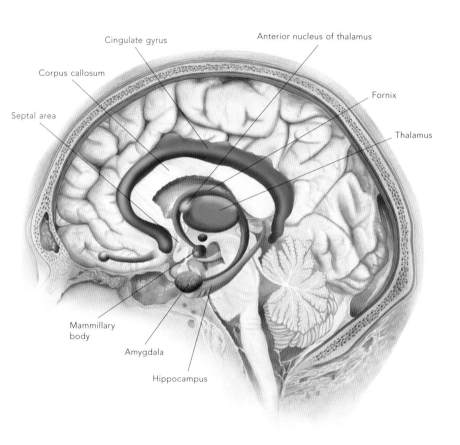

Cingulate gyrus

Anterior nucleus of thalamus

Corpus callosum

Fornix

Septal area

Thalamus

Mammillary body

Amygdala

Hippocampus

Cortical specialization & asymmetry

THE ALLOCATION OF DIFFERENT REGIONS OF THE CEREBRAL CORTEX TO HIGHER FUNCTIONS IS NOT ALWAYS SYMMETRICAL, meaning that some functions are preferentially located in one cerebral hemisphere rather than the other. A prime example of this is the positioning of the language areas (Broca's and Wernicke's areas) in the left hemisphere in 96% of right-handed people. Even the majority of left-handed people still have their language areas in the left hemisphere. In the case of Wernicke's area, this leads to anatomical asymmetry between the two sides: the upper surface of the temporal lobe is longer on the left than the right to accommodate a region within Wernicke's area that is known as the planum temporale.

Other functions that are asymmetrically placed include the preferential location of goal-directed planning in the left prefrontal cortex and the preferential localization of the rationalization of emotionally difficult decisions in the right prefrontal cortex. The right parietal cortex is also very important in making mental manipulations of objects in three-dimensional space and in judging the distances of objects. It functions as an integrative center for sensory information from the senses of touch, hearing, and vision that allows us to form a mental model of the world around us. Damage here can lead to neglect of the entire opposite side of the body and an inability to judge distances or mentally rotate objects.

▶ **FUNCTIONAL ORGANIZATION OF THE CEREBRAL CORTEX**
Some functional areas of the brain, e.g., the motor and somatosensory cortices, are symmetrical in the two hemispheres, but some functions, e.g., language and spatial perception, are preferentially located in one hemisphere rather than the other, i.e., language in the left and spatial perception in the right.

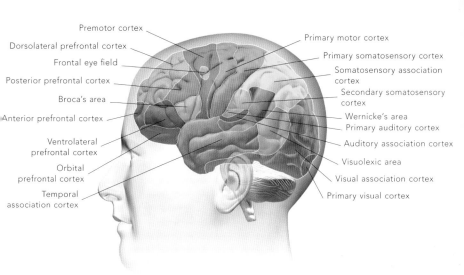

Premotor cortex
Dorsolateral prefrontal cortex
Frontal eye field
Posterior prefrontal cortex
Broca's area
Anterior prefrontal cortex
Ventrolateral prefrontal cortex
Orbital prefrontal cortex
Temporal association cortex

Primary motor cortex
Primary somatosensory cortex
Somatosensory association cortex
Secondary somatosensory cortex
Wernicke's area
Primary auditory cortex
Auditory association cortex
Visuolexic area
Visual association cortex
Primary visual cortex

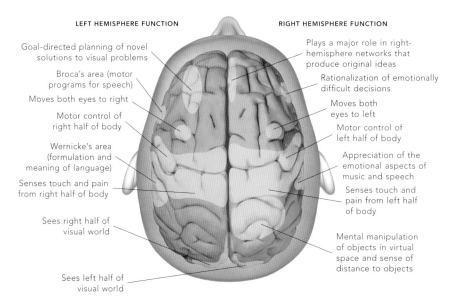

LEFT HEMISPHERE FUNCTION

Goal-directed planning of novel solutions to visual problems
Broca's area (motor programs for speech)
Moves both eyes to right
Motor control of right half of body
Wernicke's area (formulation and meaning of language)
Senses touch and pain from right half of body
Sees right half of visual world
Sees left half of visual world

RIGHT HEMISPHERE FUNCTION

Plays a major role in right-hemisphere networks that produce original ideas
Rationalization of emotionally difficult decisions
Moves both eyes to left
Motor control of left half of body
Appreciation of the emotional aspects of music and speech
Senses touch and pain from left half of body
Mental manipulation of objects in virtual space and sense of distance to objects

Language & music

IN MOST INDIVIDUALS, THE LANGUAGE CENTERS ARE LOCATED IN THE CEREBRAL CORTEX OF THE LEFT HEMISPHERE OF THE BRAIN. The two language centers traditionally identified are the motor or expressive speech area of Broca, which is located on the inferior frontal gyrus, and the receptive or sensory speech area of Wernicke, which is located at the junction of the temporal and parietal cortex, immediately posterior to the primary auditory cortex.

Patients with damage to Broca's area have a spoken language that is halting and sparse and have problems with articulation of words. Patients with lesions of Wernicke's area may have fluent speech but have a major problem with speech comprehension and significant difficulty with grammatical structure. They are also very poor at repeating speech and naming objects.

Broca's and Wernicke's areas are connected by a fiber bundle called the arcuate fasciculus, which allows streaming of information about heard speech forward toward Broca's area. This is an important function in correcting errors in our own speech. Although most language function is localized in the left hemisphere, the right hemisphere does have some capacity to understand concrete nouns, e.g., "dog" or "hammer."

The position of musical centers in the brain differs depending on the level of training. Untrained people use the right hemisphere for perception of complex sequences of sounds, such as in a melody, whereas trained musicians tend to engage the language-dominant left hemisphere more in their musical ability.

▶ LANGUAGE AREAS IN THE BRAIN

Although Broca's and Wernicke's areas (top illustration) play key roles in language, comprehension of visual language involves a pathway from the visual cortex to the visuolexic area in the inferior parietal lobule. There are also interactions between Broca's area and the facial parts of the primary motor cortex for articulation of speech.

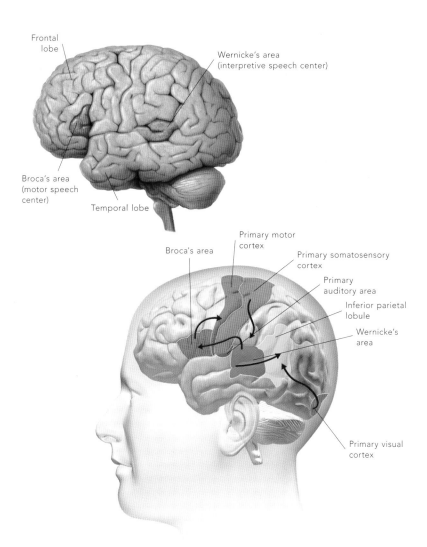

Frontal lobe

Wernicke's area (interpretive speech center)

Broca's area (motor speech center)

Temporal lobe

Broca's area

Primary motor cortex

Primary somatosensory cortex

Primary auditory area

Inferior parietal lobule

Wernicke's area

Primary visual cortex

Brain metabolism & blood flow

THE BRAIN HAS A VERY HIGH LEVEL OF METABOLIC ACTIVITY. Even though it makes up only 2% of body weight, the brain consumes 16% to 20% of the body's oxygen and nutrients. Interruption of arterial flow to the brain for as little as two minutes can cause brain damage and five minutes will cause death.

The arterial supply to the brain is by four arteries arranged in two pairs. The internal carotid arteries are branches of the common carotid arteries and provide the major source of blood for the cerebral hemispheres and front of the brain. The vertebral arteries are branches of the subclavian arteries and supply the brainstem and posterior parts of the cerebral hemisphere, e.g., the primary visual cortex. The four feeder arteries are joined together by an arterial circuit called the circle of Willis. The critical importance of the blood supply to the brain is reflected in the positioning of the arterial pressure (baro-) receptor in the carotid sinus at the origin of the internal carotid artery, and the oxygen saturation detectors in the carotid body nearby.

Arterial flow to the brain is kept fairly constant over a wide range of arterial blood pressure (about 70 to 170 mm Hg) by a process of autoregulation. Arterial flow to parts of the living brain can now be mapped with magnetic resonance imaging and positron emission scanning, allowing the identification of key functional areas for many different brain activities.

STROKE
Stroke is the transient loss of neurological function because of a vascular problem. Strokes may be due to hemorrhage (bleeding from arteries) or infarction (death of tissue due to obstruction of an artery). Contributing factors to stroke include elevated blood pressure, cigarette smoking, abnormal blood lipids, and diabetes mellitus.

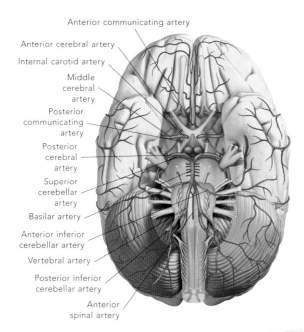

Anterior communicating artery

Anterior cerebral artery

Internal carotid artery

Middle cerebral artery

Posterior communicating artery

Posterior cerebral artery

Superior cerebellar artery

Basilar artery

Anterior inferior cerebellar artery

Vertebral artery

Posterior inferior cerebellar artery

Anterior spinal artery

▶ BLOOD SUPPLY TO THE BRAIN

The brain has a rich arterial supply (top illustration) that is fed by two internal carotid arteries and two vertebral arteries. The four arteries are joined by an arterial circle (circle of Willis) to evenly distribute arterial flow. The capillaries of the brain (lower illustration) are covered by the end-feet of astrocytes to form a barrier between the bloodstream and the tissue of the brain (blood-brain barrier).

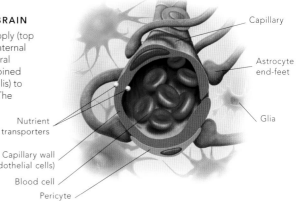

Capillary

Astrocyte end-feet

Glia

Nutrient transporters

Capillary wall (endothelial cells)

Blood cell

Pericyte

Autonomic nervous system: structure & divisions

THE AUTONOMIC NERVOUS SYSTEM IS THE PART OF THE NERVOUS SYSTEM THAT FUNCTIONS WITHOUT CONSCIOUS THOUGHT (hence, autonomously). It is traditionally divided into two divisions: the parasympathetic and the sympathetic.

The parasympathetic division is also called the craniosacral outflow because it consists of parasympathetic axons in the cranial nerves 3 (oculomotor), 7 (facial), 9 (glossopharyngeal), and 10 (vagus), as well as the pelvic splanchnic or visceral nerves from sacral segments 2 to 4. The parasympathetic outflow is active throughout most of the day, maintaining vegetative functions associated with taking on energy stores, e.g., digestion and absorption, or cleaning the airways of debris, e.g., secretion and transport of airway mucus.

The parasympathetic nervous system controls diverse functions, such as focusing the eye and constricting the pupil, contracting the urinary bladder to release urine, and penile erection.

The sympathetic nervous system is also called the thoracolumbar outflow because it consists of nerves that leave the spinal cord between the first thoracic and second lumbar spinal segments. A key component of the sympathetic nervous system is the sympathetic trunk, which is a chain of ganglia and interconnecting axons that runs from the base of the skull to the tip of the coccyx.

The sympathetic nervous system is concerned with expending energy rapidly in an emergency situation. It is most active in functions such as pupillary dilation—increasing the force, rate, and speed of cardiac contraction; elevation of blood pressure and flow to skeletal muscles; and ejaculation.

▶ **COMPONENTS OF THE AUTONOMIC NERVOUS SYSTEM**

The sympathetic nervous system (on the left) and the parasympathetic nervous system (on the right) both control many of the organs of the body.

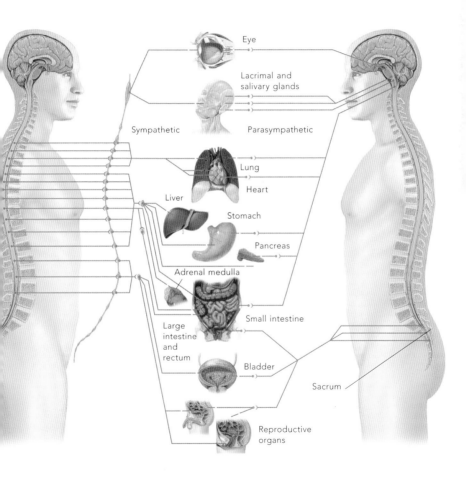

Eye

Lacrimal and
salivary glands

Sympathetic   Parasympathetic

Lung

Heart

Liver

Stomach

Pancreas

Adrenal medulla

Small intestine

Large
intestine
and
rectum

Bladder

Sacrum

Reproductive
organs

Autonomic nervous system: neurotransmitters & receptors

BOTH THE PARASYMPATHETIC AND SYMPATHETIC NERVOUS SYSTEMS CONSIST OF A SERIES OF PREGANGLIONIC AND POSTGANGLIONIC NERVE CELLS. In the parasympathetic nervous system, the preganglionic nerve cell has its cell body located in either the brainstem or the gray matter of sacral spinal segments 2 to 4. These preganglionic axons are long and terminate with a nicotinic acetylcholinergic synapse on the cell bodies of postganglionic nerve cells close to the target organ. The postganglionic parasympathetic nerve cell terminates in the target organ with a muscarinic type 2 acetylcholinergic synapse.

Preganglionic sympathetic nerve cells are located in the lateral horn of the spinal cord and have short axons. The preganglionic sympathetic outflow is identical for the control of many different organs, with preganglionic axons terminating on postganglionic nerve cells with a nicotinic acetylcholinergic synapse.

By contrast, the termination of postganglionic nerve cells on target cells may take several forms. Some sympathetic postganglionic nerve cells use norepinephrine as their neurotransmitter, acting on alpha and beta adrenoceptors. Other sympathetic postganglionic nerve cells use acetylcholine as their neurotransmitter when they contact target cells, e.g., sweat glands, with muscarinic type 2 receptors. And others contact kidney vessels use dopamine as their neurotransmitter, which binds to dopamine type 1 receptors. Finally, some postganglionic sympathetic nerve cells are modified as endocrine cells (the adrenal medulla) and directly release epinephrine and norepinephrine into the bloodstream to act on alpha and beta type adrenoceptors.

▶ **NEUROTRANSMITTERS IN THE AUTONOMIC NERVOUS SYSTEM**

Both systems use acetylcholine at the synapse between preganglionic and postganglonic cells, but the neurotransmitters used at the termination of the postganglionic cells differ significantly, particularly in the sympathetic nervous system.

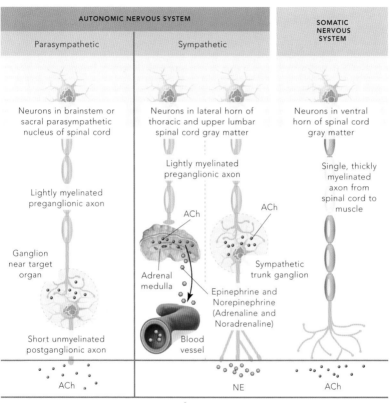

AUTONOMIC NERVOUS SYSTEM		SOMATIC NERVOUS SYSTEM
Parasympathetic	Sympathetic	

Neurons in brainstem or sacral parasympathetic nucleus of spinal cord

Neurons in lateral horn of thoracic and upper lumbar spinal cord gray matter

Neurons in ventral horn of spinal cord gray matter

Lightly myelinated preganglionic axon

Single, thickly myelinated axon from spinal cord to muscle

Lightly myelinated preganglionic axon

ACh

ACh

Ganglion near target organ

Adrenal medulla

Sympathetic trunk ganglion

Epinephrine and Norepinephrine (Adrenaline and Noradrenaline)

Short unmyelinated postganglionic axon

Blood vessel

ACh

NE

ACh

Cardiac muscle and pacemaker cells

Involuntary stimulatory or inhibitory action on visceral organs. Stimulation or inhibition depends on the neurotransmitter and receptor types in target organs

Smooth muscle and glands in gastrointestinal tract

Voluntary and stimulatory action on skeletal muscle

Enteric nervous system

THE ENTERIC NERVOUS SYSTEM IS THE COLLECTION OF NERVE CELL BODIES (ganglia) and their nerve processes (axons) that lie within the wall of the gut tube. The enteric nervous system controls the secretion of gut glands, as well as the peristalsis movements that shift food down the tube.

The enteric nervous system includes sensory neurons that detect gut distension and traction and motor nerve cells that control smooth muscle contraction. These enteric nerve cells may number as many as in the spinal cord (about 100 million) and are autonomous and self-regulating in function, but their level of activity can be influenced by the sympathetic and parasympathetic parts of the autonomic nervous system. Parasympathetic stimulation increases gut motility and secretion, whereas sympathetic stimulation decreases both.

The enteric nervous system is made up of two networks in the wall of the gut. The Meissner or submucosal plexus is located immediately under the mucosal lining of the gut. It is mainly concerned with regulation of the glands of the gut wall and with activating the submucosal smooth muscle to move the mucosa against the gut contents to facilitate absorption. Auerbach's or the myenteric plexus is located between the inner circular and outer longitudinal layers of smooth muscle of the gut tube. Its primary role is to coordinate the peristalsis that moves gut contents toward the anus.

The enteric nervous system uses many of the same neurotransmitters as the central nervous system, so drugs that act on the brain and spinal cord may also have unwanted effects on the gut. There are also many interactions between the enteric and central nervous systems.

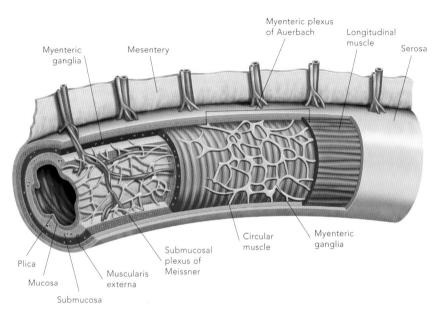

Myenteric ganglia

Mesentery

Myenteric plexus of Auerbach

Longitudinal muscle

Serosa

Circular muscle

Myenteric ganglia

Submucosal plexus of Meissner

Plica

Mucosa

Muscularis externa

Submucosa

▲ STRUCTURE OF THE ENTERIC NERVOUS SYSTEM

The enteric nervous system consists of two plexuses and their associated nerve cells. These are located in the submucosa (Meissner's plexus) and between the two layers of smooth muscle (Auerbach's plexus).

Structure of bone

BONE IS A DYNAMIC TISSUE THAT
CHANGES ITS STRUCTURE IN RESPONSE
TO PHYSICAL STRESSES AND THE
PHYSIOLOGY OF THE REST OF THE BODY.
Bone is a composite material, as it
comprises both organic and inorganic
matrices. The organic matrix, called
osteoid, contains fibers such as
collagen and is embedded within
the inorganic mineral matrix of
hydroxyapatite. Bone is a living tissue
and has cells embedded within it.

Bone can be compact and dense
(outer parts of bone) or porous and
spongy (cancellous bone around the
inner or marrow cavity). Compact
bone is formed of tightly packed units
called osteons that consist of layers
of bone (lamellae) with bone cells
(osteocytes) in lacunae arranged
around a central Haversian canal with
vessels and nerves. Volkmann's canals
running at right angles to the long axis
of the osteons connect the Haversian
canals. Spongy bone has spicules or
trabeculae of bone arranged along the
lines of force transmission. The interior
of the bone is often occupied by a
marrow cavity where fat may be stored

or blood cells manufactured, whereas
the ends of bones often participate
in joints and are thus covered with
articular cartilage to reduce friction.

The outer surface of the bone is
covered with a membrane called
periosteum that carries nerves and
vessels (periosteal arteries and veins).
Other prominences of the bone
receive the insertions of the tendons
of muscles or the attachments of
ligaments. Before adulthood, bone
grows at cartilaginous epiphyseal
plates, and remnants of these continue
into adulthood as epiphyseal lines.

▶ NAKED EYE AND MICROSCOPIC STRUCTURE OF A LONG BONE

A typical long bone has two ends, with
points of muscle attachment by tendon
insertion, as well as one or more surfaces
that are covered by articular cartilage for
participation in synovial joints. The outer
cortical part of any bone is compact and
dense to carry most of the compressive
force. Cortical bone structure has a
dense structure made of cylindrical
osteons with central Haversian canals.
The center of the bone is usually spongy
to minimize weight.

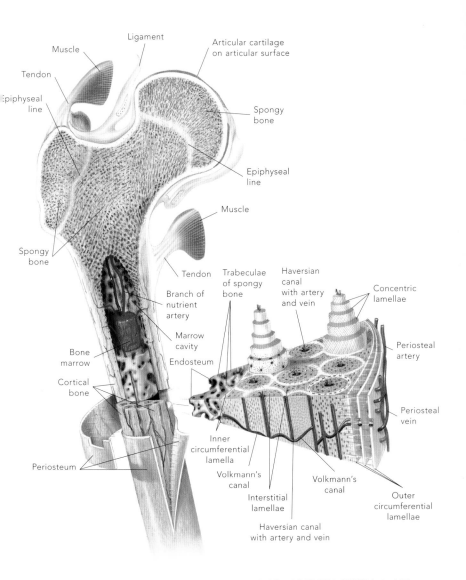

Muscle

Ligament

Articular cartilage
on articular surface

Tendon

Epiphyseal
line

Spongy
bone

Epiphyseal
line

Muscle

Spongy
bone

Tendon

Branch of
nutrient artery

Trabeculae
of spongy
bone

Haversian
canal
with artery
and vein

Concentric
lamellae

Marrow
cavity

Periosteal
artery

Bone
marrow

Endosteum

Cortical
bone

Periosteal
vein

Inner
circumferential
lamella

Volkmann's
canal

Periosteum

Volkmann's
canal

Interstitial
lamellae

Outer
circumferential
lamellae

Haversian canal
with artery and vein

Bones of the axial skeleton

THE AXIAL SKELETON IS THE PART OF THE SKELETON THAT RUNS DOWN THE MIDLINE OF THE BODY. This comprises the bones of the skull, including the cranial vault bones (frontal, parietal, temporal, occipital) protecting the brain; the bones of the skull base (sphenoid and ethmoid) beneath the brain; and the facial bones (nasal, lacrimal, maxilla, and mandible) in the front. Immediately below the mandible is a small bone called the hyoid that provides attachment for muscles of the tongue, floor of the mouth, and throat.

The skull sits upon the vertebral column, which can be divided into cervical, thoracic, lumbar, sacral, and coccygeal regions. There are seven cervical vertebrae, 12 thoracic, five lumbar, five sacral fused as a single bone, and two coccygeal. The adult human vertebral column is arranged in a series of curves to keep the head balanced above the center of gravity of the body. The curvature is concave to the back in the cervical and lumbar regions and concave to the front in the thoracic and sacral regions. The thoracic and sacral curvatures are present from embryonic life (primary curvatures), whereas the cervical and lumbar curvatures emerge later (secondary curvatures), after the infant begins to lift his or her head and stands upright, respectively.

The anterior parts of the axial skeleton in the chest include the sternum, which is divided into the manubrium, body, and xiphoid process, and the 12 pairs of ribs.

▶ **BONES OF THE SKELETON**

The bones of the body are divided into the axial skeleton (skull, hyoid, ribs, sternum, and vertebral column) and the appendicular skeleton of the four limbs. Costal cartilages ossify (turn into bone) during the fourth decade of life.

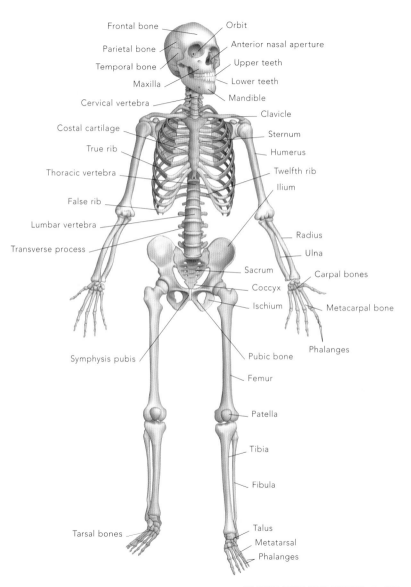

Frontal bone
Orbit
Parietal bone
Anterior nasal aperture
Temporal bone
Upper teeth
Maxilla
Lower teeth
Cervical vertebra
Mandible
Clavicle
Costal cartilage
Sternum
True rib
Humerus
Thoracic vertebra
Twelfth rib
False rib
Ilium
Lumbar vertebra
Radius
Transverse process
Ulna
Sacrum
Carpal bones
Coccyx
Ischium
Metacarpal bone
Phalanges
Symphysis pubis
Pubic bone
Femur
Patella
Tibia
Fibula
Talus
Tarsal bones
Metatarsal
Phalanges

Bones of the upper limb

THE UPPER LIMB CAN BE DIVIDED INTO TWO PARTS: the pectoral girdle, where the limb attaches to the trunk, and the upper limb proper. The bones of the pectoral girdle are the scapula and clavicle. The scapula has no direct joint with the axial skeleton but is joined to the trunk by muscle attachments. The clavicle has a single joint with the axial skeleton at the manubrium of the sternum.

The upper limb is divided into the arm—between the shoulder and elbow—with only a single bone (the humerus) and a forearm with paired bones (radius and ulna). It also includes the wrist with eight carpal bones (scaphoid, lunate, triquetrum, pisiform in the row closest to the shoulder, and trapezium, trapezoid, capitate, and hamate in the distal row); five bones of the palm of the hand (metacarpals); two finger bones (proximal and distal phalanges) in the thumb (digit 1); and three finger bones (proximal, middle, and distal phalanges) in each of digits 2 to 5.

The radius can rotate around the ulna, allowing the palm to be faced down (pronation) or up (supination). The thumb in humans is oriented to face across the palm, making opposition of the thumb tip with other digits easier and allowing humans to perform a precision grip. Additional small bones may be found within tendons; these are called sesamoid bones because of their similarity in shape to sesame seeds. They act as tiny pulleys, changing the direction of pull of tendons.

▶ **BONES OF THE UPPER LIMB FROM THE FRONT (LEFT) AND THE BACK (RIGHT)**

The bones of the upper limb consist of a pectoral girdle (scapula and clavicle), a single bone for the arm (the humerus), two bones in the forearm (radius and ulna), eight bones in the wrist (the carpals), five metacarpal bones in the palm, two phalanges in the thumb (digit 1), and three phalanges in each of digits 2 to 5.

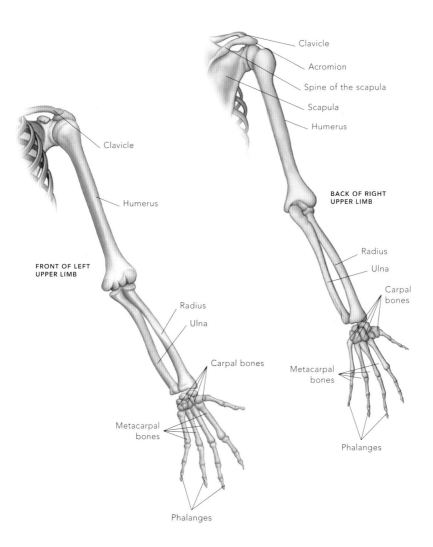

Clavicle

Acromion

Spine of the scapula

Scapula

Humerus

**BACK OF RIGHT
UPPER LIMB**

Clavicle

Humerus

**FRONT OF LEFT
UPPER LIMB**

Radius

Ulna

Radius

Ulna

Carpal bones

Carpal bones

Metacarpal
bones

Metacarpal
bones

Phalanges

Phalanges

Bones of the lower limb

THE LOWER LIMB CAN BE DIVIDED INTO TWO PARTS: the pelvic girdle, where the limb attaches to the sacrum of the vertebral column, and the lower limb proper. The pelvic girdle consists of three bones (ilium, ischium, and pubis) that are fused together at the point of the hip joint (acetabulum). The pelvic girdle provides attachments for the muscles of the lower limb and also protects the soft tissues of the pelvic cavity (urinary bladder, rectum, and reproductive organs).

The lower limb consists of the thigh, with a single bone called the femur; the leg, with two bones (the substantial tibia and the spindle-like fibula); the seven tarsal bones of the posterior foot (calcaneus, talus, navicular, cuboid, medial cuneiform, intermediate cuneiform, and lateral cuneiform); five metatarsal bones of the forefoot; and phalanges or toe bones (two in the big toe or digit 1—a proximal and a distal phalanx) and three in each of the other toes (digits 2 to 5—proximal, middle, and distal phalanges).

There is also a bone at the front of the knee—the patella—that is embedded within the tendon of the quadriceps femoris muscle of the thigh. The patella is a sesamoid bone that changes the direction of pull of the quadriceps femoris muscle when the knee is bent. Other sesamoid bones may be found embedded within the tendons at the base of the big toe.

▶ **BONES OF THE LOWER LIMB FROM BEHIND (TOP LEFT) AND THE FRONT (LOWER RIGHT)**

The bones of the lower limb include a pelvic girdle (the hip bone), the bone of the thigh (femur), the two bones of the leg (tibia and fibula), seven bones of the heel and hindfoot (the tarsals), five metatarsals of the forefoot, two phalanges in the great toe (digit 1) and three phalanges in each of toes 2 to 5.

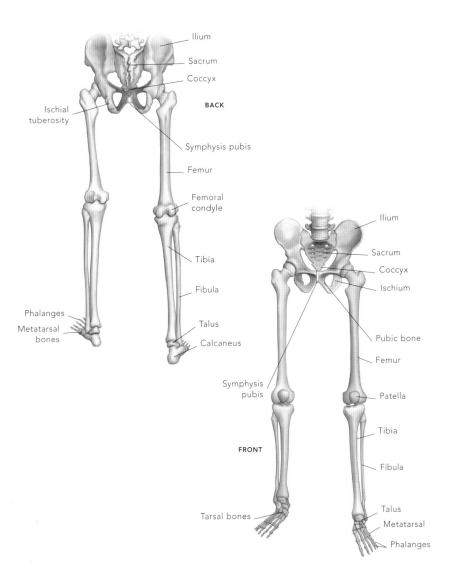

Ilium

Sacrum

Coccyx

BACK

Ischial
tuberosity

Symphysis pubis

Femur

Femoral
condyle

Tibia

Fibula

Phalanges

Metatarsal
bones

Talus

Calcaneus

Ilium

Sacrum

Coccyx

Ischium

Pubic bone

Femur

Symphysis
pubis

Patella

Tibia

Fibula

Talus

Tarsal bones

Metatarsal

FRONT

Phalanges

Joints of the body

JOINTS ARE SITES AT WHICH ONE BONE JOINS ANOTHER. Some joints can be immobile, e.g., the fibrous joints or sutures between the skull bones, or the gomphosis joints binding the teeth into the maxilla or mandible. Other joints are fibrocartilaginous, meaning that they incorporate both fibrous tissue and a cartilage disc. Fibrocartilaginous joints (symphyses) allow a greater range of movement than fibrous joints but the range is still limited. Good examples would be the intervertebral disks of the vertebral column and the pubic symphysis between the two pubic bones.

The most mobile joints are synovial joints, so called because they have a synovial fluid-filled joint cavity. The bone ends in synovial joints are coated with hyaline cartilage to minimize friction when the two bones move against each other. Synovial joints are surrounded by a synovial membrane that produces the synovial fluid as well as capsular ligaments to strengthen the entire joint. The degree of mobility at synovial joints depends on a combination of joint shape, tension

▶ **EXAMPLES OF SYNOVIAL JOINTS**

Synovial joints may be planar or gliding in type (as between many of the carpals), saddle-shaped (as between the base of the metacarpal of the thumb and the trapezium), pivot (as between the atlas and axis of the cervical vertebrae), ellipsoidal (as between the scaphoid and distal radius), hinge-shaped, e.g., the elbow, and ball-and-socket, e.g., the shoulder and hip joint.

in associated ligaments, and attachments of surrounding muscles.

Planar or gliding joints, e.g., between carpal bones, allow gliding movements of bones; hinge joints, e.g., the elbow joint, allow flexion and extension movements in one plane; pivot joints, e.g., the paired radioulnar joints, allow rotation around the long axis of the limb part; condyloid joints, e.g., at the knuckles, or ellipsoid joints, e.g., at the wrist, allow movements in two planes; and ball and socket joints, e.g., the shoulder and hip joints, allow movement in two planes plus some rotation.

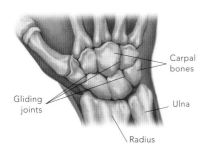

Carpal bones

Gliding joints

Ulna

Radius

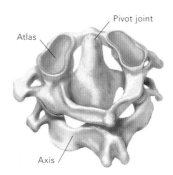

Pivot joint

Atlas

Axis

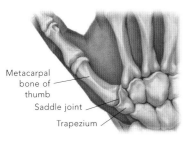

Metacarpal bone of thumb

Saddle joint

Trapezium

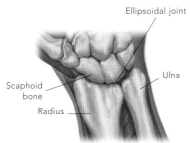

Ellipsoidal joint

Ulna

Scaphoid bone

Radius

ARTHRITIS

Arthritis literally means inflammation of a joint, but there are many types. Rheumatoid arthritis is an inflammatory connective tissue disease that affects the soft tissues of the joint (synovial membranes). Gout is an inflammatory arthritis due to the formation of crystals of uric acid in the soft tissues of the joint. Osteoarthritis is a degenerative condition of the cartilage and bone of the joint due to years of wear and tear. Instability of the joint due to damaged ligaments and weak muscles can be contributing factors.

Joints of the body *Continued*

GLIDING JOINT

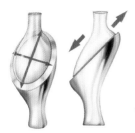

Allows sliding movement in only one plane, e.g., joints between carpal and tarsal bones.

HINGE JOINT

Allows rotational movement around a single axis so that one bone sweeps in an arc, e.g., elbow joint.

PIVOT JOINT

Allows rotation around a single axis, e.g., atlantoaxial joint (joint between the first and second cervical vertebrae).

SADDLE JOINT

Allows movement in two planes at right angles to each other but impedes rotation around the axis of the shaft, e.g., joint between base of first metacarpal and trapezium.

BALL-AND-SOCKET JOINT

Allows free movement in two planes as well as rotation around the shaft axis, e.g., glenohumeral joint of shoulder and hip joint.

ELLIPSOIDAL JOINT

Allows movement in two planes at right angles to each other but impedes rotation around the axis of the shaft, e.g., joint between scaphoid and distal radius.

► EXAMPLES OF BALL-AND-SOCKET AND HINGE JOINTS

The hip joint is a highly mechanically stable joint because of the close conformity of the ball and socket surfaces, the strength of the muscles around the joint, and the presence of a ligament that serves to anchor the head of the femur in the acetabulum.

The elbow joint is a hinge joint primarily between the trochlear notch of the proximal ulna and the trochlea of the distal humerus. It allows the ulna to sweep in an arc (flexion and extension of the elbow). The radius also articulates with the humerus, but the radiohumeral joint would be very unstable without the ulnohumeral joint.

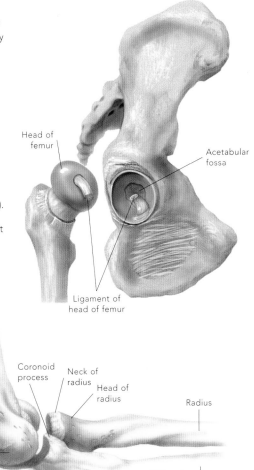

Head of femur

Acetabular fossa

Ligament of head of femur

Humerus

Coronoid process

Neck of radius

Head of radius

Radius

Medial epicondyle

Trochlea (of humerus)

Olecranon

Ulna

Voluntary muscles of the body

THE VOLUNTARY, OR SKELETAL, MUSCLES OF THE BODY USUALLY HAVE AT LEAST ONE ATTACHMENT TO THE SKELETON. Voluntary muscles should be distinguished from cardiac or smooth muscle, which are not under voluntary control and are found in the heart (cardiac muscle) or in the gut tube, urinary bladder wall, and airway walls (smooth muscle). There are about 700 voluntary muscles in the body. Some are very small, are white or light pink in color, and function in rapid twitches, e.g., the muscles that move the eyeballs. Others are large, deep red in color, and capable of sustained contraction, e.g., postural muscles of the posterior abdominal wall, such as the quadratus lumborum.

Muscles usually have Latin names that reflect their appearance. Muscles that bend a joint are called flexors, whereas those that straighten or extend a joint are called extensors. The muscles that bend the fingers are therefore called the flexor digitorum profundus or flexor digitorum superficialis. Some muscles are named according to a combination of the number of muscle heads and position. For example, the triceps brachii is the three-headed muscle of the arm, biceps brachii is the two-headed muscle of the arm, quadriceps femoris is the four-headed muscle of the thigh, and triceps surae is the three-headed muscle of the leg.

Other muscles are named according to their function. For example, the buccinator (trumpeter) is so called because it blows air out of the mouth, and the sphincter ani externus is the muscle that squeezes the anal canal. The risorius muscle of the face is so-called because it pulls the corners of the mouth backwards (*risor* is Latin for a laugher).

▶ **VOLUNTARY MUSCLES**
Voluntary muscles usually have at least one attachment to the skeleton and are under voluntary control. They have a striated appearance under the microscope due to the regular array of contractile proteins within their cytoplasm. These muscles are usually enclosed in fascial sheaths that assist the development of tension and prevent excess bulging during contraction.

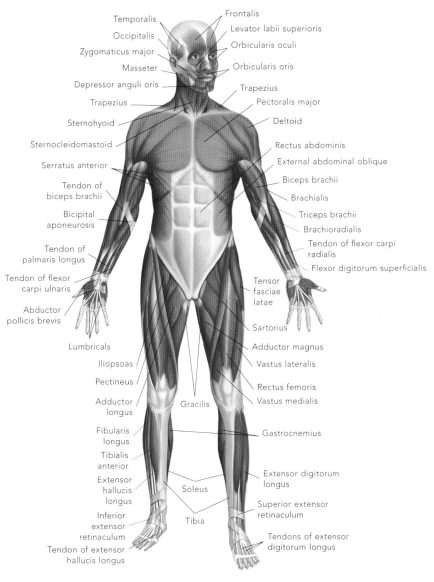

Temporalis

Occipitalis

Zygomaticus major

Masseter

Depressor anguli oris

Trapezius

Sternohyoid

Sternocleidomastoid

Serratus anterior

Tendon of
biceps brachii

Bicipital
aponeurosis

Tendon of
palmaris longus

Tendon of flexor
carpi ulnaris

Abductor
pollicis brevis

Lumbricals

Iliopsoas

Pectineus

Adductor
longus

Fibularis
longus

Tibialis
anterior

Extensor
hallucis
longus

Inferior
extensor
retinaculum

Tendon of extensor
hallucis longus

Frontalis

Levator labii superioris

Orbicularis oculi

Orbicularis oris

Trapezius

Pectoralis major

Deltoid

Rectus abdominis

External abdominal oblique

Biceps brachii

Brachialis

Triceps brachii

Brachioradialis

Tendon of flexor carpi
radialis

Flexor digitorum superficialis

Tensor
fasciae
latae

Sartorius

Adductor magnus

Vastus lateralis

Rectus femoris

Vastus medialis

Gastrocnemius

Gracilis

Extensor digitorum
longus

Soleus

Superior extensor
retinaculum

Tibia

Tendons of extensor
digitorum longus

Muscles of the head & neck

MUSCLES OF THE HEAD AND NECK INCLUDE THE FACIAL MUSCLES THAT MOVE THE SKIN OF THE FACE, muscles that move the eyes (extraocular muscles), muscles of mastication (chewing), and muscles of the neck.

The facial muscles have one attachment (the muscle origin) onto the bones of the skull and insert into the skin of the face. They are important for nonverbal communication and for protecting the openings of the eyes and mouth. Some facial muscles encircle openings in the face (the orbicularis oculi around the eye and the orbicularis oris around the mouth), but most insert around the mouth to enable the production of a range of facial expressions.

Extraocular muscles consist of a set of four arranged around the top, bottom, and side of each eyeball (the recti muscles) and obliquely oriented muscles that can rotate the eyeball. Muscles of mastication, e.g., masseter and temporalis, attach the mandible (jaw) to the skull and will pull the teeth of the mandible toward those of the maxilla during chewing.

Muscles of the neck are arranged in front of the vertebral column (prevertebral muscles) to produce flexion or bending forward of the neck, or behind the column (postvertebral muscles) to produce extension or backward bending of the neck. The sternomastoids are prominent muscles on the side of the neck that are important for maintaining the balance of the head on the neck.

▶ **MUSCLE GROUPS IN THE HEAD AND NECK**

Muscles of the head and neck include groups involved with moving the skin of the face, e.g., facial muscles like the orbicularis oculi and zygomaticus major and minor; chewing, e.g., muscles of mastication like the masseter and temporalis; and muscles that support and move the head, e.g., the sternomastoid and trapezius. Facial muscles are supplied by the branches of the facial nerve that emerge from the parotid gland.

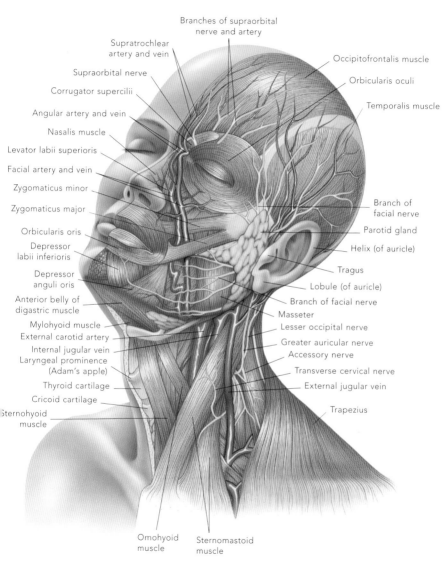

Branches of supraorbital nerve and artery

Supratrochlear artery and vein

Supraorbital nerve

Corrugator supercilii

Angular artery and vein

Nasalis muscle

Levator labii superioris

Facial artery and vein

Zygomaticus minor

Zygomaticus major

Orbicularis oris

Depressor labii inferioris

Depressor anguli oris

Anterior belly of digastric muscle

Mylohyoid muscle

External carotid artery

Internal jugular vein

Laryngeal prominence (Adam's apple)

Thyroid cartilage

Cricoid cartilage

Sternohyoid muscle

Occipitofrontalis muscle

Orbicularis oculi

Temporalis muscle

Branch of facial nerve

Parotid gland

Helix (of auricle)

Tragus

Lobule (of auricle)

Branch of facial nerve

Masseter

Lesser occipital nerve

Greater auricular nerve

Accessory nerve

Transverse cervical nerve

External jugular vein

Trapezius

Omohyoid muscle

Sternomastoid muscle

Muscles of the trunk

MUSCLES OF THE TRUNK INCLUDE THE MUSCLES OF THE CHEST AND ABDOMINAL WALLS, the muscles of the diaphragm and pelvic floor, and the muscles of the back. The muscles of the chest wall lie between the ribs (intercostal muscles) and make a contribution to inspiration and expiration of air. The muscles of the anterolateral abdominal wall consist of the rectus abdominis at the front and three layers of lateral muscles (external oblique, internal oblique, and transversus abdominis from outside to inside).

The abdominal muscles move the trunk and protect the abdominal organs from injury. They also can increase abdominal pressure in forced expiration or when expelling material from the abdominopelvic cavity, e.g., in defecation or parturition (giving birth), and when supporting the vertebral column during trunk flexion.

The muscular diaphragm between the thoracic and abdominopelvic cavities is the main muscle of lung ventilation. When it contracts, the diaphragm descends, increasing the volume of the chest cavity and drawing air into the lungs. The pelvic floor supports the pelvic organs (rectum, uterus, and urinary bladder in females; rectum, prostate, and urinary bladder in males). The pelvic floor is also important in controlling urination (micturition) and defecation.

Muscles of the back are arranged in groups in front of the vertebral column (prevertebral muscles) for flexion of the column, or in groups posterior to the vertebral column (postvertebral muscles like the erector spinae) for trunk extension.

▶ **ARRANGEMENT OF THE MUSCLES OF THE TRUNK**

Muscles of the trunk are arranged in the trunk wall (intercostal and abdominal wall muscles), between the two cavities of the trunk (the muscular diaphragm separating the thoracic and abdominopelvic cavities), and at the exit from the abdominopelvic cavity (the pelvic floor). Muscles of the vertebral column produce movement of the trunk (flexion, extension, and rotation) and support the trunk during bending.

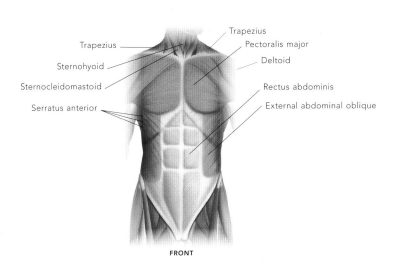

Trapezius

Trapezius

Pectoralis major

Sternohyoid

Deltoid

Sternocleidomastoid

Rectus abdominis

Serratus anterior

External abdominal oblique

FRONT

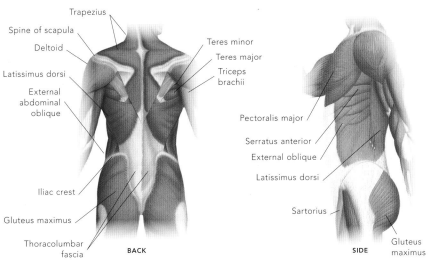

Trapezius

Spine of scapula

Deltoid

Teres minor

Teres major

Latissimus dorsi

Triceps brachii

External abdominal oblique

Iliac crest

Gluteus maximus

Thoracolumbar fascia

BACK

Pectoralis major

Serratus anterior

External oblique

Latissimus dorsi

Sartorius

Gluteus maximus

SIDE

Muscles of the upper limb

MUSCLES OF THE UPPER LIMB INCLUDE THOSE OF THE SHOULDER THAT EITHER ATTACH THE SCAPULA TO THE TRUNK (serratus anterior, rhomboideus major and minor, and levator scapulae) or insert onto the upper humerus (pectoralis major, deltoid, teres major, and latissimus dorsi; and the rotator cuff muscles, such as supraspinatus, infraspinatus, teres minor, and subscapularis).

The muscles of the arm can be divided into the flexors of the elbow (biceps brachii and brachialis) on the front of the arm and the extensor of the elbow (triceps brachii) on the back of the arm. The muscles of the forearm are concerned with rotation of the radius around the ulna, e.g., pronator teres and supinator; flexion or extension of the wrist, e.g., flexors carpi radialis and ulnaris, and extensors carpi radialis and ulnaris; or movement of the digits, e.g., flexor digitorum superficialis and profundus, flexor pollicis longus, and extensor digitorum.

Muscles within the hand are found in the fleshy thenar eminence at the base of the thumb (short flexor,

opposer, and abductor of the thumb) or in the hypothenar eminence at the base of the little finger (flexor, opposer, and abductor of the little finger). Some muscles are found between the metacarpal bones (interossei and lumbricals). The interosseus muscles separate the digits or bring them together, whereas the lumbricals assist with the flexion of the knuckles when the wrist is extended. The flexion of the knuckles with the wrist extended is important for precision grip, e.g., threading a needle.

▶ **MUSCLES OF THE UPPER LIMB FROM THE FRONT AND BACK**

Muscles of the upper limb are grouped into shoulder, arm, forearm, and hand muscles. Powerful movements of the fingers and thumb are driven by muscles of the forearm, with tendons running into the hand to insert onto the phalanges, whereas precision movements depend more on the intrinsic muscles of the hand in the thenar, hypothenar, and palmar compartments.

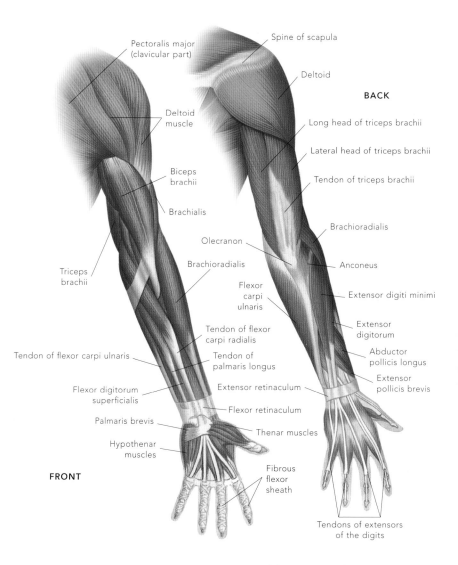

Pectoralis major
(clavicular part)

Spine of scapula

Deltoid

BACK

Deltoid
muscle

Long head of triceps brachii

Lateral head of triceps brachii

Biceps
brachii

Tendon of triceps brachii

Brachialis

Brachioradialis

Olecranon

Triceps
brachii

Anconeus

Brachioradialis

Extensor digiti minimi

Flexor
carpi
ulnaris

Extensor
digitorum

Tendon of flexor
carpi radialis

Abductor
pollicis longus

Tendon of flexor carpi ulnaris

Tendon of
palmaris longus

Extensor
pollicis brevis

Flexor digitorum
superficialis

Extensor retinaculum

Flexor retinaculum

Palmaris brevis

Thenar muscles

Hypothenar
muscles

Fibrous
flexor
sheath

FRONT

Tendons of extensors
of the digits

Muscles of the lower limb

MUSCLES OF THE LOWER LIMB ARE DIVIDED INTO BUTTOCK (GLUTEAL), THIGH, LEG, AND FOOT MUSCLES. Gluteal muscles include the very large gluteus maximus that extends the thigh when climbing stairs and the smaller gluteus medius and gluteus minimus that are critically important for supporting the hip when the opposite leg is being swung forward during the process of walking.

The muscles of the thigh can be divided into three parts. The anterior compartment, with the large quadriceps femoris (made up of rectus femoris, vastus medialis, vastus lateralis, and vastus intermedius) extends the knee. The medial compartment of the thigh contains the adductor muscles (gracilis, adductor magnus, adductor brevis, and adductor longus) that bring the femur toward the midline. The posterior compartment of the thigh contains the hamstring muscles (semitendinosus, semimembranosus, and biceps femoris) that extend at the hip and flex at the knee.

The leg is divided into posterior, lateral, and anterior compartments. The posterior compartment is occupied by the gastrocnemius and soleus muscles that flex the ankle (a movement called plantarflexion), as well as the long flexors of the toes. The anterior compartment contains muscles that extend the toes and dorsiflex the foot (tip big toe upward). The lateral compartment contains the peroneal or fibular muscles (fibularis longus, fibularis brevis, and fibularis tertius) that bend the foot outward (eversion). The muscles of the foot are arranged in four layers and move the toes.

▶ **MUSCLES OF LOWER LIMB FROM THE FRONT AND BACK**

The muscles of the thigh and leg are large and powerful. The muscles of the foot are arranged in four layers but do not have the sophisticated level of control that the muscles of the hand do.

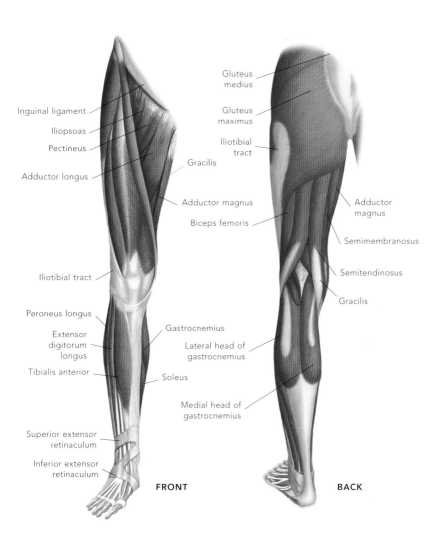

Gluteus medius

Inguinal ligament

Iliopsoas

Pectineus

Adductor longus

Gracilis

Gluteus maximus

Iliotibial tract

Adductor magnus

Biceps femoris

Adductor magnus

Semimembranosus

Semitendinosus

Gracilis

Iliotibial tract

Peroneus longus

Extensor digitorum longus

Tibialis anterior

Gastrocnemius

Lateral head of gastrocnemius

Soleus

Medial head of gastrocnemius

Superior extensor retinaculum

Inferior extensor retinaculum

FRONT

BACK

Neuromuscular junctions

THE NEUROMUSCULAR JUNCTION IS THE SITE AT WHICH THE AXON OF A MOTOR NEURON CONTACTS THE MUSCLE FIBER THAT IT INNERVATES. The motor neuron axon has a swelling at its end called the axon terminal or synaptic bulb, which contains synaptic vesicles filled with the neurotransmitter acetylcholine (ACh).

When an action potential reaches the axon terminal, synaptic vesicles move to the axonal membrane and molecules of acetylcholine are released into the space (the synaptic cleft) between the axon terminal and the sarcolemma (the muscle membrane). These molecules bind to ACh receptors (ligand-gated sodium ion channels) on the specialized region of the sarcolemma called the motor endplate. The binding activates sodium channels, allowing sodium ions to enter the cell to produce local depolarization of the sarcolemma (end-plate potential) beneath the neuromuscular junction.

Usually multiple end-plate potentials need to be generated before functional contraction of the muscle occurs. The ACh molecules are degraded in the synaptic cleft by an enzyme called acetylcholinesterase, so the motor axon has to keep firing continuously in order to release more ACh to maintain muscle contraction.

Botulinum toxin binds irreversibly to the axon terminals of motor neurons and blocks release of ACh from synaptic vesicle, so it can be used in very small quantities to relieve chronic muscle spasm. Nerve gases and some insecticides block the enzyme acetylcholinesterase, keeping the neuromuscular junction flooded with ACh and causing muscle twitching.

▶ **NEUROMUSCULAR JUNCTION STRUCTURE**

The neuromuscular junction consists of an axon terminal branch and its bulbs and the membrane of the muscle being contacted. Packets of acetylcholine (ACh) in synaptic vesicles are released into the cleft between the two membranes. The molecules of acetylcholine attach to receptors in the muscle membrane and activate the opening of sodium channels to produce an end-plate potential in the muscle.

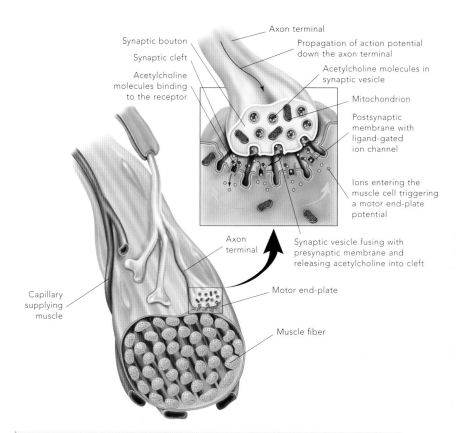

Axon terminal

Synaptic bouton

Synaptic cleft

Acetylcholine molecules binding to the receptor

Propagation of action potential down the axon terminal

Acetylcholine molecules in synaptic vesicle

Mitochondrion

Postsynaptic membrane with ligand-gated ion channel

Ions entering the muscle cell triggering a motor end-plate potential

Axon terminal

Synaptic vesicle fusing with presynaptic membrane and releasing acetylcholine into cleft

Capillary supplying muscle

Motor end-plate

Muscle fiber

MYASTHENIA GRAVIS

Myasthenia gravis is an autoimmune condition in which the body attacks the ACh receptors in the neuromuscular junction. Symptoms include general muscular weakness, particularly affecting the eyes, arms, head, and chest. The condition is more common in women than men, and the age of onset is also much younger for women than men. Treatment is by acetylcholinesterase inhibitors, e.g., neostigmine, to make more ACh available at the neuromuscular junction.

Structure of muscle

SKELETAL MUSCLES ARE COMPOSED OF MUSCLE FIBER CELLS, which are elongated cells that contain the actin and myosin contractile proteins. Each muscle cell has a plasma membrane called the sarcolemma, surrounding a cytoplasm containing myofibrils (muscle protein bundles). The sarcolemma extends as transverse tubules (or T-tubules) deep into the interior of the cell to surround each myofibril. The muscle cells have multiple nuclei arranged around the periphery of the cell. Within the cell is a system of tubules called the sarcoplasmic reticulum. This is a modified smooth endoplasmic reticulum that serves to store calcium ions and release these onto the contractile proteins when muscle contraction is initiated (see pp. 174–175).

The striated appearance of skeletal muscle is due to the regular arrangement of contractile proteins into functional units called sarcomeres. Two of the major types of filaments within the sarcomeres are the thick filaments, composed of myosin, and the thin filaments, composed of actin, as well as regulatory proteins (troponin and tropomyosin). Elastic filaments are also present and contribute to mechanical energy storage. The contraction of muscle depends on the transient binding of heads of myosin proteins in the thick filaments with the actin filaments, accompanied by a shifting or sliding of the thick filaments relative to the thin ones (the sliding filament mechanism of contraction).

MUSCULAR DYSTROPHY

Muscular dystrophy is a genetic condition in which there are defects in muscle proteins and death of muscle cells. Some types of muscular dystrophy (Duchenne and Becker) are caused by a mutation on the X chromosome and therefore primarily affect boys (who have only one X chromosome), because girls with the faulty gene will often have a functional gene on their other X chromosome. Most boys with Duchenne muscular dystrophy are unable to walk by puberty and often die before the age of 45 years.

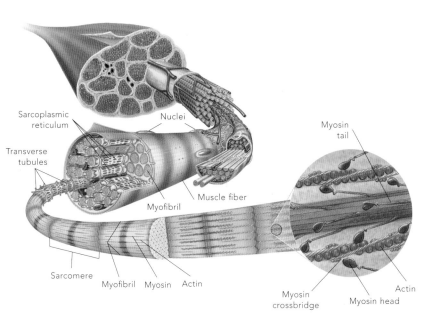

Sarcoplasmic reticulum

Nuclei

Myosin tail

Transverse tubules

Muscle fiber

Myofibril

Myosin crossbridge

Myosin head

Actin

Sarcomere

Myofibril

Myosin

Actin

▲ **MICROSCOPIC STRUCTURE OF MUSCLE**

Skeletal muscle has a complex structure at microscopic and ultrastructural levels. Each muscle is composed of elongated cells called muscle fibers that are arranged in parallel. Each muscle fiber has regular arrays of contractile proteins arranged in a series of contractile elements called sarcomeres. The interdigitation of the myosin and actin filaments allows tension to be developed by the adenosine triphosphate (ATP)-powered interaction of the two filament types.

Actin, myosin, troponin, & tropomyosin

THE TWO KEY CONTRACTILE PROTEINS IN THE MUSCLE CELL ARE ACTIN AND MYOSIN. The myosin protein chains are grouped together as thick filaments, which consist of a central segment of myosin molecule tails and two ends made up of clustered myosin heads. The actin protein chains form the bulk of the thin filaments of the muscle. Actin is a bead-shaped protein that has an active site for interaction with myosin heads. Actin forms two intertwined chains within the thin filaments.

These actin chains are accompanied in the thin filament by paired chains of the protein tropomyosin and occasional beads of the protein troponin. Tropomyosin is a regulatory protein that covers the active sites of actin when the muscle is at rest. Troponin is another regulatory protein that holds the tropomyosin in place.

Together the troponin and tropomyosin molecules help to initiate and halt contraction. Elastic filaments are coiled protein molecules called titin that run through the coil of the thick filaments and help to stabilize them. Elastic filaments resist overstretching

of muscle and provide elasticity, so the muscle tends to spring back to its original length after it has been stretched.

▶ **MICROSTRUCTURE OF MUSCLE**

Muscle tissue has regular arrays of contractile, regulatory, and elastic proteins. Contractile proteins, e.g., actin and myosin, provide the actual contractile force. Regulatory proteins, e.g., troponin and tropomyosin, ensure that contraction occurs only when the appropriate stimulation has been received. Elastic proteins store mechanical potential energy between contractions and smooth out the effects of individual twitches.

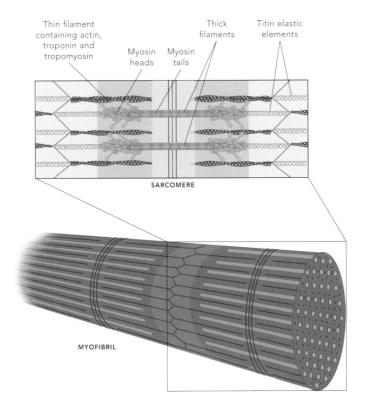

Thin filament containing actin, troponin and tropomyosin

Myosin heads

Myosin tails

Thick filaments

Titin elastic elements

SARCOMERE

MYOFIBRIL

Excitation-contraction coupling

WHEN AN ACTION POTENTIAL REACHES THE AXON TERMINAL OF THE MOTOR NEURON, acetylcholine is released into the synaptic cleft of the motor end plate, causing a motor muscle action potential to spread along the muscle fiber. The T-tubule system of the muscle fiber allows this action potential to spread into the interior of the muscle fiber, carrying the signal to the vicinity of all myofibrils.

The action potential triggers the release of calcium ions stored in the sarcoplasmic reticulum, and these ions quickly move to the sarcomeres (see p.170), where the calcium binds to the troponin molecule of the thin filaments, changing its shape. The change in shape of the troponin molecule removes the blocking action of the other regulatory protein, tropomyosin, so that the active sites on the actin filament become exposed. The heavy myosin heads are now able to bind to the actin. The physical attachment of the myosin to the actin also causes the bending of the myosin toward the center of the sarcomere, pulling the heavy filaments along the thin filaments and shortening the sarcomere segment. This binding and pulling expends energy in the form of adenosine triphosphate (ATP).

Once the action potential ends, calcium ions are removed from the troponin and reabsorbed into the sarcoplasmic reticulum in preparation for the next contraction. Tropomyosin blockage of the actin active sites is restored and the contraction ceases. This entire cycle lasts only a few tens of milliseconds.

▶ **THE STEPS IN EXCITATION-CONTRACTION COUPLING**

There are a series of steps in the process of excitation-contraction coupling. They lead from an action potential of the muscle membrane, through release of calcium ions (Ca^{2+}) into the vicinity of the contractile proteins to bind to the troponin (1), binding and bending of myosin to pull actin filaments toward the center of the sarcomere (2), binding of ATP and release of the actin-myosin cross-bridges (3), and cocking of the myosin head ready for the next contraction (4). As part of the process of contraction, ATP (adenosine triphosphate) is converted to ADP (adenosine diphosphate) and inorganic phosphate.

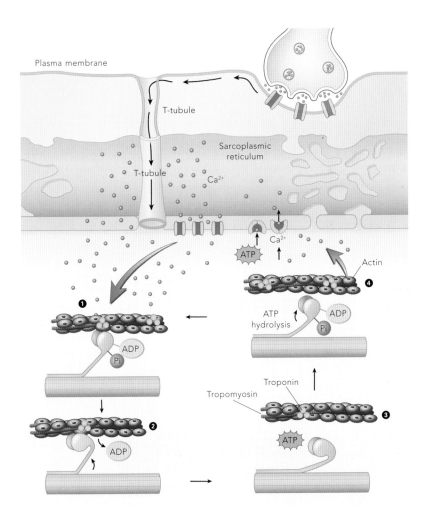

Plasma membrane

T-tubule

T-tubule

Sarcoplasmic reticulum

Ca^{2+}

ATP

Ca^{2+}

Actin

❶

ADP

Pi

❷

ADP

ATP hydrolysis

ADP

Pi

Troponin

Tropomyosin

❸

ATP

❹

Muscle tension & length

THE AMOUNT OF TENSION THAT CAN
BE DEVELOPED IN A MUSCLE IS CLOSELY
RELATED TO THE LENGTH OF THE
MUSCLE FIBER. Therefore, it is
dependent on the microstructure of
the muscle tissue. The most important
determinant of muscle tension
achieved with each contraction is
the number of actin/myosin binding
sites that are activated in the muscle
microsegments or sarcomeres. This
in turn is dependent on the relative
overlap of thin and thick filaments in
the sliding filament mechanism of
muscle contraction.

When the muscle is short, some of
the actin thin filament is aligned with
a part of the thick filament where
there are few myosin heads, so the
actin/myosin interaction and the
tension developed are low (see top
illustration). As the muscle is
stretched slightly, more myosin heads
can be aligned with the active actin
sites so more tension is developed. If
the stretching process continues, the
actin filaments are aligned with only
the very ends of the thick filaments,
the potential for actin/myosin binding

is limited, and the tension developed
with any contraction is low.

This relationship at the microscopic
level is also manifested at the naked eye
level because each muscle is composed
of millions of sarcomeres arranged in
a series. So any muscle will have an
optimal muscle length at which a given
contraction will produce the highest
tension, based on the extent of the
actin/myosin alignment in each of its
component sarcomeres. Remember
that muscles are attached to bones and
produce movements at joints, so the
relationship also carries over to joint
angle. There is an optimal joint angle at
which a muscle is at its most effective.
In the case of the biceps brachii flexing
the elbow, this optimal zone is in the
90° to 120° range (see lower illustration).

▶ **THE RELATIONSHIP BETWEEN
SARCOMERE LENGTH AND
TENSION AND BETWEEN JOINT
ANGLE AND FORCE**

The number of myosin heads that can
interact with actin filaments is a key
determinant of the muscle force that
can be developed in the sarcomere.
This is optimal at a sarcomere length
a little beyond resting length.

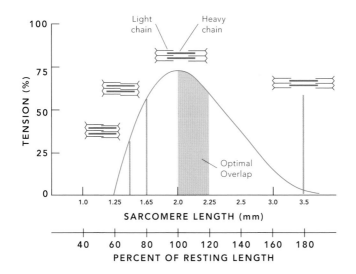

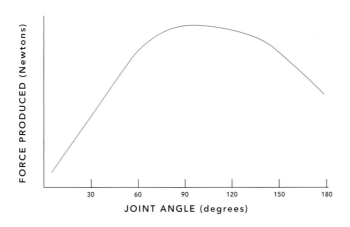

Summation of contraction & tetany

VERY FEW MUSCLES CAN PRODUCE A USEFUL CONTRACTION WITH A SINGLE ACTIVATION, i.e., a single twitch (see graph a). The only exception to this would be the eye flicks produced by the extraocular muscles, where the eye is quickly moved to a new position. For almost all muscle activation, and certainly for any sustained contraction of postural muscles, many individual muscle twitches must be combined smoothly in a process that is called temporal summation.

When muscle fibers are activated at a frequency of more than 8 Hz, i.e., 8 cycles per second, the tension developed with each contraction is not immediately dissipated but is stored in the elastic elements of the muscle (see graph b). These are found within the muscle fiber itself, i.e., the elastic filaments of the sarcomere, and also within the connective tissue attachments of the muscle, i.e., tendons.

This storage of elastic potential energy allows the tension developed with a series of muscle twitches to be much greater than that achievable with a single twitch. From about 10 to 30 Hz,

the individual peaks of each muscle contraction are still visible on the trace of tension versus time (unfused tetany, see graph c), but once about 40 Hz is reached, the overall tension in the muscle is many times higher than that for the individual twitch, and the trace of tension versus time is said to exhibit fused tetany. In fused tetany the level of tension developed is a smooth plateau (see graph d).

▶ **MUSCLE TWITCHES TO PRODUCE SMOOTH CONTRACTION**

Individual muscle twitches (top) last only a few hundred milliseconds, so the tension developed by many twitches must be added together to achieve smooth contraction of muscle. Temporal summation of twitches is achieved when the frequency of muscle stimulation rises above 8 Hz (bottom). Note that when stimulation frequency rises to 10 Hz or more, the tension developed by the muscle is also much greater than single twitches.

MUSCLE TWITCH

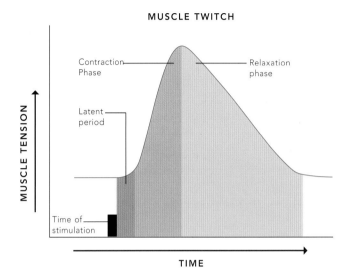

Contraction Phase

Relaxation phase

Latent period

MUSCLE TENSION

Time of stimulation

TIME

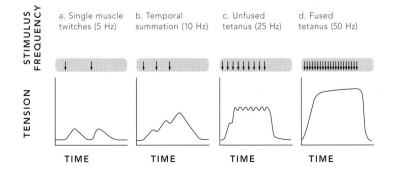

STIMULUS FREQUENCY

a. Single muscle twitches (5 Hz)

b. Temporal summation (10 Hz)

c. Unfused tetanus (25 Hz)

d. Fused tetanus (50 Hz)

TENSION

TIME

TIME

TIME

TIME

Motor unit recruitment

A MOTOR UNIT IS A MOTOR NERVE CELL AND ALL THE MUSCLE FIBERS THAT ITS AXONS CONTACT AND CAN ACTIVATE. In the same way as many muscle twitches must be combined to produce a smooth movement, the activation of many motor nerve cells and the muscle fibers they control, i.e., many motor units, makes a key contribution to sustained and effective muscle contraction. The total contractile force developed by the combined action of a group of motor units within a skeletal muscle is always greater than that developed by any one of its motor units. The contraction is also smoother and more sustained when many motor units are involved.

Some muscles are controlled by a large number of motor nerve cells, i.e., there are many motor units for a given muscle mass. This is beneficial when fine control of muscle fiber activation is required for precision movement, e.g., for eye movements, or for the fine control of the fingers, e.g., when playing a piano.

Other muscles have a relatively small number of motor units per muscle mass, meaning that a single motor nerve cell controls many muscle fibers. This is seen where power is more important than precision, e.g., in postural muscles of the back or large powerful muscles of the lower limb like the quadriceps femoris at the front of the thigh.

▶ **RECRUITMENT OF MULTIPLE MOTOR UNITS TO INCREASE MUSCLE TENSION**

The muscle fibers supplied by a single motor nerve cell make up a motor unit. The tension developed in a muscle during a movement can be gradually raised by increasing the number of motor units activated. Some muscles have only a few muscle fibers per motor nerve cell and have fine motor control, e.g., extraocular muscles, whereas others have many muscle fibers supplied by a single motor nerve cell, e.g., thigh muscles.

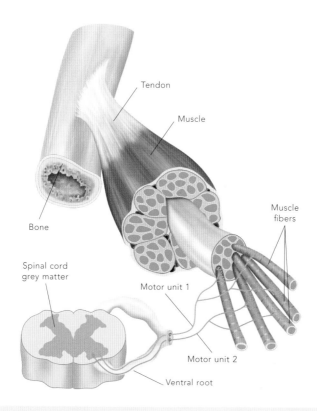

Tendon

Muscle

Muscle fibers

Bone

Spinal cord grey matter

Motor unit 1

Motor unit 2

Ventral root

MOTOR NEURON DISEASE

Motor neuron disease is a condition of progressive degeneration of motor nerve cells in the spinal cord, brainstem, and/or motor areas of the cerebral cortex. Symptoms usually begin after the age of 40 and are more common in men than women. Symptoms and signs include difficulty performing voluntary movement, muscle weakness, and muscle wasting or rigidity. Sensory function is unaffected. The cause is poorly understood, but some cases are genetic.

Smooth muscle

SMOOTH MUSCLE CELLS ARE VERY IMPORTANT COMPONENTS OF INTERNAL ORGANS LIKE THE GUT AND BLOOD VESSELS. Unlike striated skeletal muscle, smooth muscle cells are flattened with a single ovoid nucleus in the center of the cell. Smooth muscle gets its name because it is unstriped, meaning that it does not contain the regular arrays of contractile proteins that give both skeletal and cardiac muscle their striped appearance.

Nevertheless, smooth muscle cells contain the contractile proteins actin and myosin just like other muscle cells. The key difference is that the contractile proteins are not regularly arranged but form a network within the cytoplasm. Actin filaments in smooth muscle cells are arranged obliquely and anchored to proteins called dense bodies. Most smooth muscle cells are also linked to each other by specialized regions of the cell membrane called gap junctions. These allow the transmission of electrical impulses between smooth muscle cells that help coordinate their function when they are grouped in large numbers, e.g., in the walls of major arteries or the gut tube.

Unlike skeletal muscle, smooth muscle can be stimulated to contract by many influences apart from nerve contacts. These include mechanical stretch and hormones. Some smooth muscles cells serve as pacemakers, providing rhythmicity to the contraction of the entire surrounding smooth muscle population. These pacemakers are especially important in the uterus during labor, where they coordinate a wave of contraction that proceeds from the top of the uterus toward the cervix.

▶ **SMOOTH MUSCLE STRUCTURE AND FUNCTION**

Smooth muscle cells differ from both cardiac and skeletal muscle cells in a number of significant ways. Smooth muscle cells have a central nucleus, unlike skeletal muscle cells. The actin and myosin are not arranged in regular arrays, so the smooth muscle cells do not have a striped appearance as cardiac and skeletal muscle cells do. Smooth muscle cells contract in a sustained fashion rather than in twitches and are under autonomic control, acting either as multi- or single-unit groups.

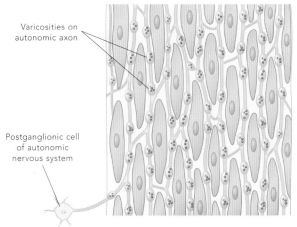

Varicosities on autonomic axon

Postganglionic cell of autonomic nervous system

MULTI-UNIT SMOOTH MUSCLE SYSTEM (IRIS OF EYE, PARTS OF MALE REPRODUCTIVE TRACT, VESSEL WALLS, ARRECTOR PILI MUSCLES OF SKIN)

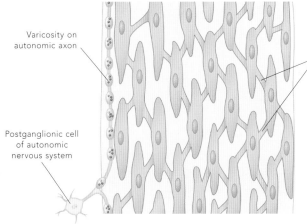

Varicosity on autonomic axon

Gap junctions electrically connecting smooth muscle cells into a single functional unit

Postganglionic cell of autonomic nervous system

SINGLE-UNIT (VISCERAL) SMOOTH MUSCLE SYSTEM (WALL OF DIGESTIVE TRACT, GALL BLADDER, URINARY BLADDER)

Overview of the circulatory system

THE CIRCULATORY SYSTEM IS DIVIDED INTO THE PULMONARY AND SYSTEMIC CIRCULATIONS. The pulmonary circulation carries deoxygenated blood from the right ventricle to the capillary beds of the lungs through the pulmonary trunk and its branches. The oxygenated blood from the lungs is returned to the left side of the heart (left atrium) through the tributaries of the four pulmonary veins.

The systemic circulation carries oxygenated blood from the left ventricle through the aorta and its branches to supply the capillary beds of the body tissues with oxygen and nutrients. Venous blood from the systemic capillaries travels in the tributaries of either the superior vena cava (from the upper limb and head) or the inferior vena cava (from the rest of the body) to enter the right atrium, where the cycle begins anew.

Major systemic arteries include the common carotid arteries in the head and their branches, supplying the face and scalp (external carotid arteries) or the brain (internal carotid arteries). The upper limb is supplied by the subclavian artery, which becomes the axillary and then the brachial artery before branching just above the elbow into radial and ulnar arteries. The descending aorta gives branches to the gut tube (celiac, superior mesenteric, and inferior mesenteric arteries) and the kidneys (renal arteries) before branching into common iliac arteries just above the pelvis. The common iliac arteries give rise to the internal iliac arteries (for supply of pelvic viscera) and the external iliac arteries (for supply of the lower limb and lower abdominal wall). The external iliac artery becomes the femoral artery when it enters the lower limb. The femoral artery becomes the popliteal artery behind the knee before giving off anterior and posterior tibial arteries to supply the leg and foot.

▶ **SYSTEMIC ARTERIES AND VEINS OF THE BODY**

Systemic arteries are all branches of the aorta that leaves the left ventricle of the heart. Systemic venous return is via the superior and inferior vena cavae into the right atrium.

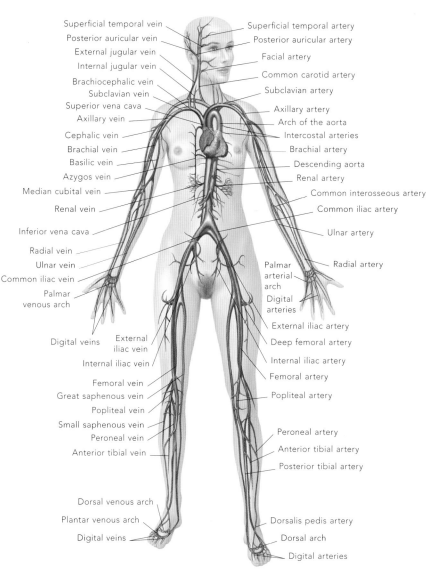

Superficial temporal vein
Posterior auricular vein
External jugular vein
Internal jugular vein
Brachiocephalic vein
Subclavian vein
Superior vena cava
Axillary vein
Cephalic vein
Brachial vein
Basilic vein
Azygos vein
Median cubital vein
Renal vein
Inferior vena cava
Radial vein
Ulnar vein
Common iliac vein
Palmar venous arch
Digital veins
External iliac vein
Internal iliac vein
Femoral vein
Great saphenous vein
Popliteal vein
Small saphenous vein
Peroneal vein
Anterior tibial vein
Dorsal venous arch
Plantar venous arch
Digital veins

Superficial temporal artery
Posterior auricular artery
Facial artery
Common carotid artery
Subclavian artery
Axillary artery
Arch of the aorta
Intercostal arteries
Brachial artery
Descending aorta
Renal artery
Common interosseous artery
Common iliac artery
Ulnar artery
Palmar arterial arch
Digital arteries
Radial artery
External iliac artery
Deep femoral artery
Internal iliac artery
Femoral artery
Popliteal artery
Peroneal artery
Anterior tibial artery
Posterior tibial artery
Dorsalis pedis artery
Dorsal arch
Digital arteries

Basic structure of the heart & cardiac muscle

THE HEART IS A FOUR-CHAMBERED PUMP WITH CARDIAC MUSCLE IN ITS WALLS. The two atria (left and right) are relatively thin-walled, lower-pressure chambers and receive blood from the pulmonary and systemic veins, respectively.

The main pumping chambers of the heart are the two ventricles (left and right). The left ventricle is the thickest-walled chamber because it must reach pressures of at least 120 mm Hg during contraction. It receives oxygenated blood from the left atrium through the mitral or bicuspid valve (left atrioventricular valve) and pumps that blood to the body through the aortic valve and aorta. The right ventricle receives blood from the right atrium through the tricuspid valve (right atrioventricular valve) and pumps that blood to the lungs through the pulmonary valve and pulmonary trunk. Its wall is about one-third the thickness of the left ventricle because the pressure developed in the right ventricle during contraction (about 25 mm Hg) is much lower than that in the left ventricle.

The bulk of the mass of the heart is muscle called the myocardium, with cardiac muscle fibers electrically joined to each other by specialized zones of cell membrane called intercalated disks. These electrical connections between muscle cells ensure that the atrial and ventricular muscle fibers each act as an electrical syncytium or single electrical unit, so when a muscle action potential is generated in one part of the atria or ventricles, it will spread through the entire connected cardiac muscle. The atrial and ventricular syncytia are insulated from each other by the connective tissue of the cardiac fibrous skeleton so that atrial contraction can be kept separate from ventricular contraction.

▶ **STRUCTURE OF THE HEART**

An external view of the anterior surface of the heart (top illustration) is opened out in the lower illustration, which shows the chambers and great vessels. Note the greater thickness of the left ventricle compared to the right ventricle.

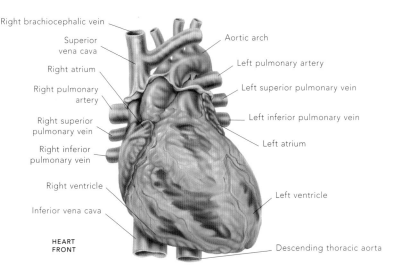

Right brachiocephalic vein
Superior vena cava
Right atrium
Right pulmonary artery
Right superior pulmonary vein
Right inferior pulmonary vein
Right ventricle
Inferior vena cava

Aortic arch
Left pulmonary artery
Left superior pulmonary vein
Left inferior pulmonary vein
Left atrium
Left ventricle
Descending thoracic aorta

HEART FRONT

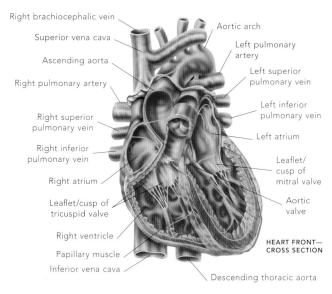

Right brachiocephalic vein
Superior vena cava
Ascending aorta
Right pulmonary artery
Right superior pulmonary vein
Right inferior pulmonary vein
Right atrium
Leaflet/cusp of tricuspid valve
Right ventricle
Papillary muscle
Inferior vena cava

Aortic arch
Left pulmonary artery
Left superior pulmonary vein
Left inferior pulmonary vein
Left atrium
Leaflet/cusp of mitral valve
Aortic valve
Descending thoracic aorta

HEART FRONT—CROSS SECTION

Heart development & fetal circulation

THE EMBRYONIC HEART DEVELOPS AS A TUBULAR STRUCTURE WITH VENOUS AND ARTERIAL ENDS. The tube folds in the middle and develops internal walls to produce the four-chambered heart of the adult. The function of the lungs is rather different in a fetus than in a newborn, and this is reflected in the changes that must occur in the circulation around the time of birth.

In a fetus, all oxygenation of blood is performed by the placenta. Oxygenated blood returns from the placenta by the left umbilical vein and bypasses the liver by flowing through the ductus venosus to reach the inferior vena cava. The fetal lungs are uninflated, so oxygenated blood from the placenta is diverted across the wall between the two atria (the interatrial septum) through an opening known as the foramen ovale. Once the oxygenated blood has reached the left atrium, it can be distributed to the body tissues by the left ventricle and aorta. Any residual oxygenated blood that does enter the right ventricle can be shunted from the pulmonary trunk to the aorta across the ductus arteriosus.

At birth, the lungs inflate and peripheral resistance drops in the capillary bed of the lungs. The pressure rises in the left atrium and drops in the right atrium, closing the flap of the foramen ovale and diverting venous blood from the inferior vena cava into the right ventricle to be pumped toward the lungs. Flow drops in the ductus venosus and the ductus arteriosus, which then become fibrosed remnants still visible in adult life.

▶ **FETAL AND NEONATAL CIRCULATION**

Significant changes must occur at birth to transform the fetal circulation to the adult arrangement. The foramen ovale between the two atria will shut, and the ductus arteriosus and ductus venosus will close. These changes direct blood from the systemic veins to the lungs for oxygenation.

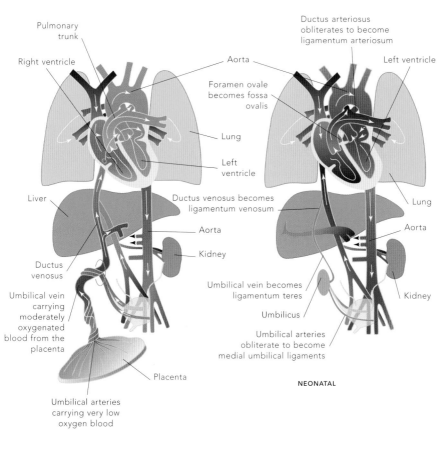

Pulmonary trunk

Right ventricle

Aorta

Foramen ovale becomes fossa ovalis

Left ventricle

Ductus arteriosus obliterates to become ligamentum arteriosum

Left ventricle

Lung

Left ventricle

Liver

Ductus venosus becomes ligamentum venosum

Aorta

Kidney

Lung

Aorta

Kidney

Ductus venosus

Umbilical vein carrying moderately oxygenated blood from the placenta

Umbilical vein becomes ligamentum teres

Umbilicus

Umbilical arteries obliterate to become medial umbilical ligaments

Placenta

NEONATAL

Umbilical arteries carrying very low oxygen blood

FETAL

Cardiac cycle & function of heart valves

THE HEART BEATS APPROXIMATELY 70 TIMES PER MINUTE AT REST, and each heartbeat represents the completion of one cardiac cycle. The cycle begins with flow of venous blood into the two atria and then into the ventricles from the respective circulations. When the atria are full, an impulse is initiated at the sinoatrial node in the right atrium, which activates contraction of the two atria. This contraction pushes the blood through the atrioventricular valves (tricuspid and mitral valves) into the respective ventricles.

In the meantime, the electrical impulse has traveled from the right atrium through the connective tissue wall between the atria and ventricles to reach the specialized conducting tissue known as the atrioventricular bundle. Activation spreads from here through the bundle branches to the ventricular musculature, inducing ventricular depolarization and contraction (ventricular systole).

When pressure in the ventricle rises above that in the atria, the tricuspid and mitral valves close, generating the first heart sound. As the pressure begins to rise in the ventricles, the aortic and pulmonary valves open, allowing the ejection of blood from the ventricles into the aorta and pulmonary trunk, respectively. Eventually, the ventricular muscle ceases contraction and begins to relax (ventricular diastole).

When the pressures in the ventricles fall below that in the aorta and pulmonary valves, the aortic and pulmonary valves close, producing the second heart sound. As ventricular pressure continues to drop, it will fall below that in the respective atrium, and the atrioventricular valves open to allow blood to flow into the relaxed ventricles, beginning the cycle again.

▶ **EVENTS IN THE CARDIAC CYCLE**

The cardiac cycle is the sequence of chamber filling and contraction that occurs with every heartbeat. It starts with venous return to the atria and ventricles (top left), followed by atrial contraction to prime the ventricles (top right), followed by ventricular contraction to force blood into the aorta and pulmonary trunk (bottom left), and finally ventricular relaxation so that venous return can fill the atria and ventricles again (bottom right).

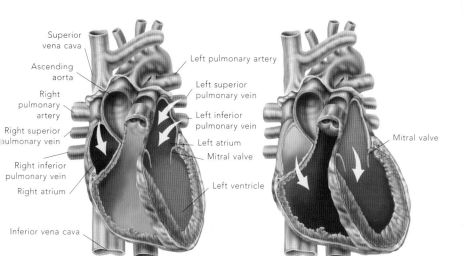

Superior vena cava
Ascending aorta
Right pulmonary artery
Right superior pulmonary vein
Right inferior pulmonary vein
Right atrium
Inferior vena cava

Left pulmonary artery
Left superior pulmonary vein
Left inferior pulmonary vein
Left atrium
Mitral valve
Left ventricle

1. VENOUS BLOOD RETURNS TO ATRIA

Mitral valve

2. ATRIAL CONTRACTION TO PUMP-PRIME VENTRICLES

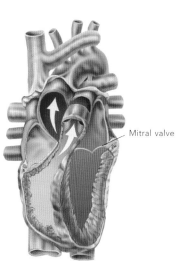

Mitral valve

3. VENTRICULAR SYSTOLE (CONTRACTION) EXPELS BLOOD FROM HEART

Mitral valve

4. VENTRICLES RELAX AND VENOUS BLOOD RETURNS TO ATRIA

Action potential of cardiac muscle

THE HEART MUSCLE, OR MYOCARDIUM, HAS CONTRACTION ACTION POTENTIALS JUST LIKE SKELETAL MUSCLE, but there are some special features that reflect the critical role that the cardiac muscle plays throughout life. These special features ensure that cardiac contraction can be sustained for a longer duration, e.g., more than 300 milliseconds, than skeletal muscle, without the requirement of repeated nerve stimulation. It is also critical that cardiac muscle has a relatively prolonged refractory period, i.e., it cannot be restimulated, so that the risk of potentially fatal cardiac fibrillation is minimized.

The resting membrane potential of cardiac muscle is about –85 mV (millivolts), and the cardiac muscle action potential reaches a membrane potential of +20 mV because of activation of voltage-gated ion channels in the sarcolemma.

The cardiac muscle action potential has four phases. The first is a rapid depolarization phase due to the activation and opening of voltage-gated sodium (Na^+) channels. The second stage is an initial repolarization phase due to the abrupt inactivation of the sodium channels. The third phase is a distinctive plateau phase, which is due to the slow opening of calcium ion (Ca^{2+}) channels. This plateau phase is important because it provides sustained contraction and a very prolonged refractory period that permits the heart rate to be kept slow enough to pump blood effectively. If the plateau phase did not exist, the heart rate would be as much as 15 times higher, and it would be difficult to achieve adequate filling of the ventricles before the next contraction began.

The final phase is a repolarization phase, when both the sodium and calcium ion channels return to normal and potassium channels allow the movement of potassium out of the muscle cell, making the membrane ready for the next activation by the pacemaking tissue.

STRUCTURAL FEATURES OF CARDIAC MUSCLE

Cardiac muscle has some special structural and functional features. Cardiac muscle cells are connected by special junctions called intercalated disks that turn the atria and ventricles into electrically connected units called syncytia. Cardiac muscle cells also have very prolonged action potentials with long refractory periods that reduce the risk of fibrillation. The vertical scale on the action potential trace above is the membrane potential in millivolts. Resting membrane potential is -85 to 90 mV and the action potential spike can reach +10 to 20 mV.

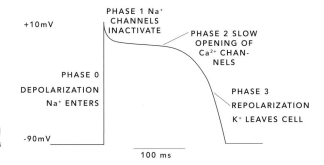

+10mV

PHASE 1 Na⁺ CHANNELS INACTIVATE

PHASE 2 SLOW OPENING OF Ca^{2+} CHANNELS

PHASE 0 DEPOLARIZATION Na⁺ ENTERS

PHASE 3 REPOLARIZATION K⁺ LEAVES CELL

-90mV

100 ms

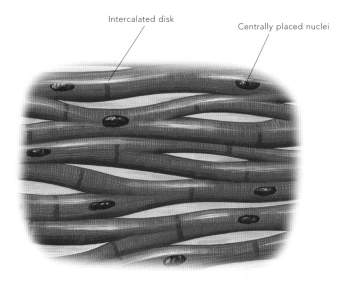

Intercalated disk

Centrally placed nuclei

Excitation-contraction coupling in cardiac muscle

THE PROCESS OF EXCITATION-CONTRACTION COUPLING IN CARDIAC MUSCLE CELLS IS SIMILAR TO THAT SEEN IN SKELETAL MUSCLE, but with some features associated with the sustained duration of the plateau phase of the cardiac muscle action potential. The signal to contract is also most likely to come from an adjacent cell rather than an axon contacting a motor end-plate because all cardiac muscle cells are electrically connected into syncytia by tight junctions between the cells called intercalated disks.

Spread of an action potential to the cardiac muscle cell from an adjacent muscle cell induces the opening of voltage-gated calcium channels so that calcium enters the cell. This in turn induces calcium release from intracellular stores in the sarcoplasmic reticulum through the action of calcium on ryanodine receptor channels. The flooding of the myofibrils with intracellular calcium brings about the necessary transformation of the regulatory proteins troponin and tropomyosin to unlock the active sites of the actin thin filaments. The light actin binds to the myosin, much as in skeletal muscle, and contraction begins.

The prolonged calcium influx and the sustained release of intracellular calcium stores keep the actin active sites exposed for longer and ensure that the contraction is much more sustained than with skeletal muscle. Relaxation of cardiac muscle is just as important as contraction and is an energy-requiring process that uses ATP. It occurs when calcium is returned to intracellular stores and the ions detach from the troponin regulatory proteins. Calcium also must be returned to the outside of the cell by sodium/calcium exchangers.

▶ **FUNCTIONAL FEATURES OF CARDIAC MUSCLE**
The special functional features of cardiac muscle are due to differences in the ion channels. In particular, the plateau phase is due to the prolonged opening of the calcium channels that allows calcium to enter the cell as potassium exits.

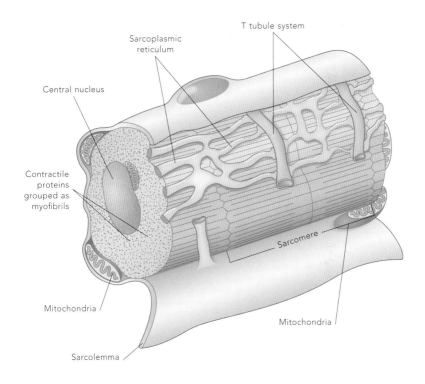

T tubule system

Sarcoplasmic
reticulum

Central nucleus

Contractile
proteins
grouped as
myofibrils

Sarcomere

Mitochondria

Mitochondria

Sarcolemma

Frank-Starling law & cardiac output

THE FRANK-STARLING LAW DESCRIBES THE RELATIONSHIP BETWEEN PRELOAD (the amount of venous blood that flows into a ventricle) and the stroke volume (the amount of blood that is expelled from the ventricle during systole). For a normal heart at rest, stroke volume is about 2.4 fl oz (70 ml). The resting heart beats at about 70 times per minute, so the cardiac output per minute is approximately 10 pints (5 liters). The more that ventricle muscle is stretched by infilling with blood during ventricular diastole, the stronger the contraction of the cardiac muscle and the more blood that will be pumped out.

The underlying mechanism for this is the degree of overlap between the actin and myosin filaments in the sarcomeres of the cardiac muscle cells. Stretching of the heart muscle cells leads to a better overlap between the actin and myosin, i.e., there are more heavy myosin heads in contact with more active sites on the actin filaments, and the heart muscle is able to achieve more contractile force.

It can be seen from the diagram that there is a limit to the relationship between preload and stroke volume. If the heart muscle is stretched too much, the actin and myosin filaments slide too far apart and the number of points of interaction between the two begins to drop. This actually leads to a drop in stroke volume, and the heart is said to be decompensating. This situation arises in chronic heart failure when the heart becomes overdilated and globular in shape.

CHRONIC HEART FAILURE

Chronic heart failure is the gradual decline in cardiac function, usually the result of multiple small incidents of cardiac muscle death (infarctions). The gradual loss of cardiac muscle and its replacement with fibrous tissue make the heart flabby and dilated. On chest X-rays the heart will look globular and enlarged. The patient will have difficulty performing even modest exercise and will show the effects of accumulation of fluid in the systemic or pulmonary veins (leg swelling and breathlessness).

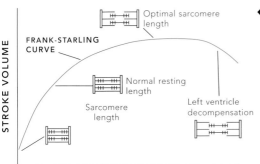

FRANK-STARLING CURVE

Optimal sarcomere length

Normal resting length

Sarcomere length

Left ventricle decompensation

STROKE VOLUME

VENTRICULAR END-DIASTOLIC VOLUME

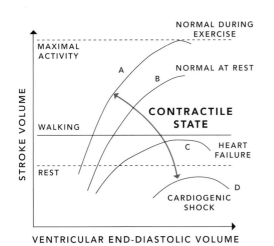

NORMAL DURING EXERCISE

MAXIMAL ACTIVITY

A

B

NORMAL AT REST

CONTRACTILE STATE

WALKING

C

HEART FAILURE

REST

D

CARDIOGENIC SHOCK

STROKE VOLUME

VENTRICULAR END-DIASTOLIC VOLUME

◀ **THE RELATIONSHIP BETWEEN STROKE VOLUME AND END-DIASTOLIC VOLUME**

The Frank-Starling curve describes the relationship between the degree of stretching of cardiac muscle fibers at the end of ventricular rest (ventricular end-diastolic volume, or VEDV) and the power developed in subsequent contraction (as shown by the stroke volume). There is an optimal VEDV that is determined by the number of actin-myosin interactions in the muscle (top illustration). In exercise with sympathetic stimulation (A), the curve may move up and to the left, allowing increased contractility (bottom illustration). In disease states, e.g., heart failure (C), and eventually cardiogenic shock (D), the curve drops to the right so that there is not enough stroke volume for even modest exercise (walking) or just to survive with the body at rest in bed.

Impulse conduction in the heart

HEART MUSCLE GENERATES ELECTRICAL ACTIVITY DURING EXCITATION-CONTRACTION COUPLING. The heart also contains conducting and pacemaking tissues that set the normal heart rhythm and spread electrical impulses through the heart to signal the heart muscle to contract. The electrical activity generated by both the conducting/pacemaking tissue and the heart muscle tissue can be detected by electrodes attached to the skin (see pp. 200–201).

The most important pacemaking tissue is the sinoatrial node (SA node), which is located in the right atrium near the entry of the superior vena cava. The SA node sets the pace for heart rhythm and has an intrinsic rate of approximately 80 depolarizations per minute, although in the intact heart this is reduced to about 70 per minute by the influence of the vagus nerve.

Other important conducting/pacemaking tissues are the atrioventricular node, which is found in the right atrium just above the tricuspid valve, and the atrioventricular bundle, which pierces the connective tissue wall that electrically isolates the atria from the ventricles.

One particularly important function of the atrioventricular bundle is that it slows the rate of impulse conduction from the atria to the ventricles so that the atria have time to contract and prime the ventricles with blood before ventricular contraction (ventricular systole) begins. After they have reached the ventricles, electrical impulses spread through the two ventricles by the respective left and right bundle branches and their Purkinje fiber branches. These pathways allow all muscle tissue throughout the ventricles to be activated simultaneously for a coordinated contraction.

▶ **PACEMAKING AND IMPULSE CONDUCTION IN THE HEART**

The primary pacemaker of the heart is the sinoatrial node in the right atrium. From here, the impulse travels through the atrial muscle to the atrioventricular node, into the ventricles by the atrioventricular bundle, and down the respective bundle branches to reach all the muscle cells of the ventricles.

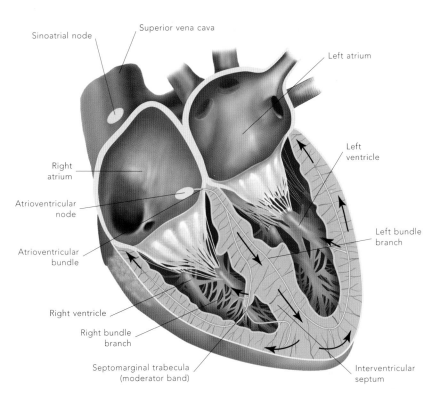

Sinoatrial node

Superior vena cava

Left atrium

Right atrium

Atrioventricular node

Atrioventricular bundle

Left ventricle

Left bundle branch

Right ventricle

Right bundle branch

Septomarginal trabecula (moderator band)

Interventricular septum

ARRHYTHMIA

Arrhythmias are abnormalities of the heart rhythm, usually due to problems with the conducting and pacemaking tissues of the heart or abnormal electrical activity of cardiac muscle. Arrhythmias of the atria are generally less serious than those of the ventricle, but atrial fibrillation can cause the formation of life-threatening thrombi (blood coagulation) that can block arteries in the brain or limbs. Ventricular fibrillation is fatal without cardioversion (application of an electrical shock to return the heart to normal rhythm).

Electrocardiograms (ECGs)

THE ELECTRICAL ACTIVITY OF THE HEART CAN BE STUDIED USING AN ELECTROCARDIOGRAM (ECG), which relies on electrodes attached to the limbs and chest wall to detect waves of electrical activity moving across the heart when it beats. The standard ECG electrodes are attached to the left arm, left leg, and right arm in a configuration called Einthoven's triangle. There are also usually six electrodes (numbered V1 to V6) attached across the chest, specifically to analyze impulses from the ventricles.

The ECG shows peaks in electrical activity that are related to electrical events in the different parts of the heart, specifically the depolarization and repolarization of the atria and the ventricles. For example, there is a distinct wave at the point of atrial depolarization (the P wave), another when the ventricles depolarize (the QRS complex), and another when the ventricles repolarize (the T wave). No discrete wave can be seen for atrial repolarization because the large QRS complex obscures it.

The ECG gives useful information on the cardiac electrical rhythm (both atrial and ventricular because these can be separated in heart block) and can detect a variety of abnormal rhythms (dysrhythmias), such as atrial or ventricular fibrillation, paroxysmal atrial tachycardia, and ventricular tachycardia. The ECG also allows the analysis of the spread of electrical activity through the ventricles. Electrical conduction through the ventricles can be disordered if there is damage to the heart muscle, e.g., after death of cardiac muscle from myocardial infarction.

▶ **ELECTRICAL EVENTS IN THE CARDIAC CYCLE AND THE ECG**

Electrical activity in the cardiac cycle begins with atrial depolarization (P wave on the ECG). When the impulse reaches the ventricles and ventricular depolarization occurs, a QRS complex is generated on the ECG. This is a long and complex wave because of the time it takes for the impulse to spread throughout the large ventricular wall. Atrial repolarization doesn't produce a visible wave on the ECG because it occurs at the same time as the large QRS complex. Finally, ventricular repolarization produces a T wave on the ECG at the time the ventricles relax.

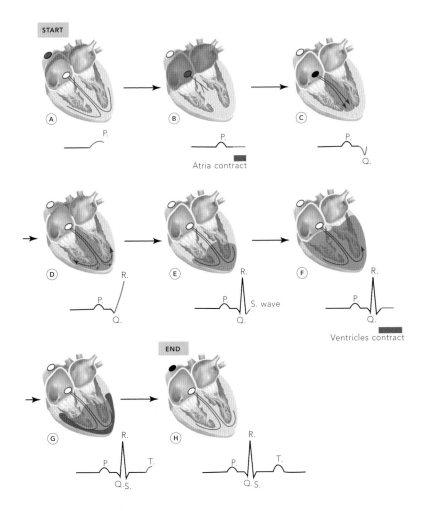

START

A
P.

B
P.
Atria contract

C
P.
Q.

D
R.
P.
Q.

E
R.
P.
Q.
S. wave

F
R.
P.
Q.
Ventricles contract

G
R.
P.
T.
Q. S.

END

H
R.
P.
T.
Q. S.

Physics of blood flow: Poiseuille's law

THE BLOOD VESSELS OF THE BODY ARE ESSENTIALLY TUBES OR PIPES, and the behavior of blood as it flows through them is the same as for any fluid in a pipe. The law describing this relationship is called Poiseuille's law. This law states that the volume rate of flow, i.e., milliliters (ml) per minute, of blood through an artery is directly proportional to the pressure drop down the tube, multiplied by the constant pi and the radius of the vessel raised to the fourth power.

Volume rate of flow is inversely proportional to the viscosity of the fluid and the length of the tubing. Note that blood flows much faster in the center of a blood vessel than close to the vessel wall (a feature of a Newtonian fluid) because of friction with the lining of the vessel.

Note also that the radius of any blood vessel is a critically important factor in determining the volume rate of flow of the fluid. Just halving the radius of a vessel, e.g., when the arterial wall is narrowed by a disease called atherosclerosis, can reduce the volume rate of flow through that artery by an

▶ **EFFECTS OF VESSEL INTERNAL RADIUS ON VOLUME FLOW**
A small change in the internal radius (r) of a blood vessel can have profound effects on the rate of flow because volume rate of flow (in ml per minute) is a function of the internal radius raised to the fourth power. Doubling the internal radius of a blood vessel can increase blood flow 16-fold. ΔP is the change in pressure down the length of the vessel.

astonishing 16-fold. Changes in the viscosity of the blood—e.g., when the number of red blood cells in the blood is increased, leukaemic cells or myeloma proteins are present, or the patient's blood has an increased lipid (fat) content—may also significantly reduce blood flow.

THE r⁴ FACTOR

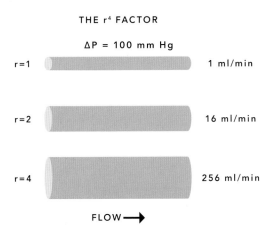

$\Delta P = 100$ mm Hg

r = 1   1 ml/min

r = 2   16 ml/min

r = 4   256 ml/min

FLOW ⟶

ATHEROSCLEROSIS

Atherosclerosis is a condition in which fatty-fibrous plaques develop in the walls of arteries. These plaques can narrow the internal radius of the artery, causing significant reductions in blood flow. The surface of the plaque may also slough off, leading to the exposure of collagen and triggering the formation of thrombi (coagulated blood), which can in turn also cause arterial blockage. The development of atherosclerosis has several risk factors, some of which are not preventable—e.g., genes, age, and male gender—whereas others can be modified, e.g., excess blood lipids, high blood pressure, cigarette smoking, a sedentary lifestyle, obesity, and diabetes mellitus. Atherosclerosis can affect the coronary arteries of the heart, causing chest pain on exertion (angina pectoris), a heart attack (myocardial infarction), or chronic loss of muscle tissue (chronic cardiac failure). If it affects the arteries to the brain or eye, stroke and blindness can result. If it affects the lower limbs, the toes may become gangrenous. Atherosclerosis in penile arteries causes erectile dysfunction.

Arterial pressure & its measurement

PRESSURE IN THE VESSELS OF THE CIRCULATION CHANGES SUBSTANTIALLY ALONG THE CIRCUIT. Close to the heart in the aorta, the pressure will be pulsatile, oscillating between systolic and diastolic pressures (usually 120 and 80 mm Hg, respectively), due to the rhythmic pumping of the heart. By the time the smaller arterioles in the tissue have been reached, the pressure oscillations have been dampened and the mean pressure is beginning to drop. Pressure in the capillaries varies from about 25 mm Hg at the arterial end to about 15 mm Hg at the venous end. Pressure in medium to large veins is about 10 mm Hg, but this can be significantly affected by hydrostatic pressure when standing.

Arterial blood pressure is the product of the amount of blood being pumped into the vascular tree every minute (the cardiac output) and the resistance to flow in the peripheral circulation (peripheral resistance). Cardiac output is in turn dependent on the heart rate and the volume of blood ejected with each stroke (stroke volume). Peripheral resistance can be influenced by secreted factors in the blood (humoral factors like norepinephrine), the control exerted by the sympathetic nervous system on smooth muscle of the small arterioles, and local factors like the capacity of tissues to regulate their regional flow with diffusible chemicals.

Blood pressure is measured by a sphygmomanometer, which compresses the brachial artery until flow is stopped before gradually lowering the cuff pressure and listening to changes in the sounds heard in the depression in front of the elbow (the cubital fossa).

▶ **PRESSURE CHANGE IN THE SYSTEMIC VASCULAR TREE**

Pressure is high and pulsatile in the aorta and other arteries with a high content of elastic tissue. Pressure drops rapidly when the terminal arterioles are reached and then declines slowly through the capillary bed and venous side of the circulation.

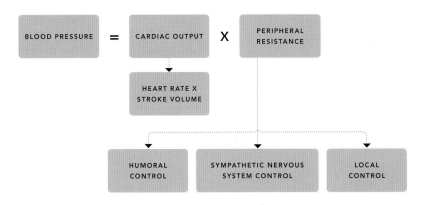

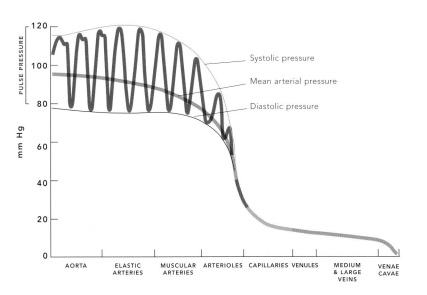

Baroreceptors & reflexes

IT IS VITALLY IMPORTANT THAT THE ARTERIAL PRESSURE IS KEPT WITHIN A NARROW OPTIMAL RANGE. This is to ensure adequate perfusion of vital organs like the brain without causing vessel rupture. The two main sensors of arterial pressure are located in the arch of the aorta and in a dilated part of the beginning of the internal carotid artery (the carotid sinus). In both cases, these pressure sensors, or baroreceptors, detect the stretching of the arterial wall and signal this information to the brain by cranial nerve 9 (the glossopharyngeal nerve) for information from the carotid sinuses or by cranial nerve 10 (the vagus nerve) for the aortic arch baroreceptor.

Control from the nervous system to regulate blood pressure may run in sympathetic nerves or by circulating factors (see pp. 208–209), but because of the required timescale, the rapid response to a rise in blood pressure must rely on nervous control. In this feedback circuit, a rise in arterial pressure is detected by the baroreceptors, and the information is transmitted to a control center in the medulla oblongata. This leads to a decrease in sympathetic outflow and a rise in paraysmpathetic outflow.

Through a combination of relaxation of arteriolar smooth muscle, decreased ventricular contractility, and slowing of the heart rate, the two key determinants of blood pressure (namely peripheral resistance and cardiac output) are reduced. The drop in arterial blood pressure feeds back to decrease the sensory stimulation of the baroreceptors.

▶ THE RESPONSE TO ELEVATED BLOOD PRESSURE

The response to a rise in blood pressure requires sensors to detect the pressure rise, control centers in the brainstem to analyze and determine the necessary response, and actions through the autonomic nervous system to change peripheral resistance and cardiac output. The neurotransmitter norepinephrine acts on different receptor types—α on arteriole smooth muscle and β_1 on the ventricular muscle and the pacemaking tissue of the sinoatrial node.

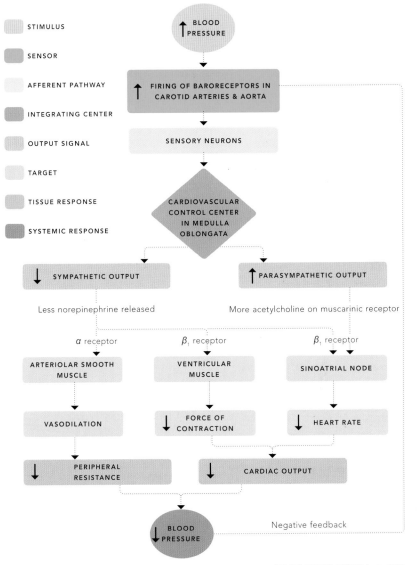

STIMULUS

SENSOR

AFFERENT PATHWAY

INTEGRATING CENTER

OUTPUT SIGNAL

TARGET

TISSUE RESPONSE

SYSTEMIC RESPONSE

↑ BLOOD PRESSURE

↑ FIRING OF BARORECEPTORS IN CAROTID ARTERIES & AORTA

SENSORY NEURONS

CARDIOVASCULAR CONTROL CENTER IN MEDULLA OBLONGATA

↓ SYMPATHETIC OUTPUT

↑ PARASYMPATHETIC OUTPUT

Less norepinephrine released

More acetylcholine on muscarinic receptor

α receptor

β_1 receptor

β_1 receptor

ARTERIOLAR SMOOTH MUSCLE

VENTRICULAR MUSCLE

SINOATRIAL NODE

VASODILATION

↓ FORCE OF CONTRACTION

↓ HEART RATE

↓ PERIPHERAL RESISTANCE

↓ CARDIAC OUTPUT

↓ BLOOD PRESSURE

Negative feedback

Regulation of arterial pressure

SEVERAL FACTORS DETERMINE ARTERIAL PRESSURE. The total blood volume is a determinant of both cardiac output and the volume of the vascular tree. Cardiac output itself is a critical determinant of arterial pressure and depends on the product of stroke volume and heart rate. Peripheral resistance to arterial flow is determined by the relative diameter of the arterioles that lead into the capillary beds. When these are opened by relaxation of the precapillary sphincter smooth muscle, peripheral resistance is lowered. The relative volumes of blood in the arterial versus venous compartments of the circulation can also contribute to arterial pressure. When arterial blood volume is high, cardiac output and blood pressure are also high.

The long-term and ongoing regulation of arterial pressure depends on negative feedback loops that involve humoral factors, such as the renin angiotensin system and atrial natriuretic peptide. Atrial natriuretic protein is released from the atrium when blood pressure rises. It will cause vasodilation and increased excretion of sodium and water at the kidney, returning the blood pressure to normotension. When blood pressure drops, the kidney releases renin, which catalyzes the conversion of angiotensinogen to angiotensin I, which is in turn converted to angiotensin II. The latter promotes reabsorption of sodium and water from the urine, vasoconstriction of the peripheral arterioles, and a return to normal pressure.

▶ **FACTORS DETERMINING ARTERIAL BLOOD PRESSURE**

Arterial pressure depends on a combination of factors that fall into humoral (hormonal) and neural (autonomic) categories. Hormonal regulation of blood pressure, e.g., by atrial natriuretic protein and the renin-angiotensin system, acts over a longer time frame (hours to days) than autonomic (seconds to minutes), but both work together throughout life.

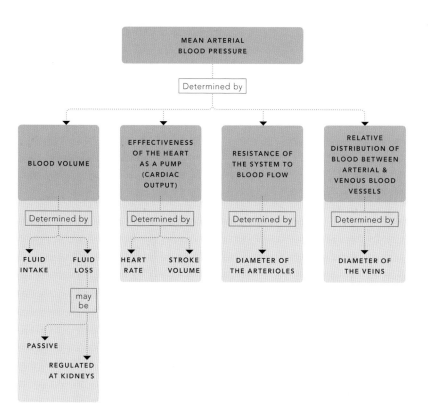

Control of cardiac output

CARDIAC OUTPUT IS A FUNCTION OF BOTH THE AMOUNT OF BLOOD EJECTED FROM THE HEART WITH EACH BEAT (stroke volume, or SV) and the heart rate (HR). Normal cardiac output is 10.5 pints (5 liters) per minute and is the product of a stroke volume of 2.4 fl oz (70 ml) and a heart rate of 70 beats per minute.

Heart rate is determined primarily by the rate of spontaneous depolarization of the conducting and pacemaking tissue of the sinoatrial node. Sympathetic stimulation increases this rate, whereas parasympathetic stimulation lowers it.

Stroke volume is determined by the force of contraction of the cardiac muscle (ventricular myocardium). By the Frank-Starling law (see pp. 196–197), contractility of the cardiac muscle is closely determined by the end-diastolic volume, i.e., the amount of blood within the ventricles immediately before contraction begins, because this governs the length of the cardiac muscle cells just before they begin contraction.

End-diastolic volume in turn depends on the amount of venous return to the heart, which is affected by the muscle pumps of the lower limbs and the respiratory pump of the diaphragm. Sympathetic innervation and circulating levels of epinephrine (released from the adrenal medulla) can also affect the cardiac contractility and the state of venous constriction.

HYPERTENSION

Hypertension is elevated arterial blood pressure in the systemic circulation. This means a resting systolic pressure above 140 mm Hg and a resting diastolic pressure above 90 mm Hg. About 90% of cases of hypertension are due to lifestyle and dietary factors (excess weight, excess salt, smoking, and alcohol). The rest are due to identified disease such as renal or endocrine disorders.

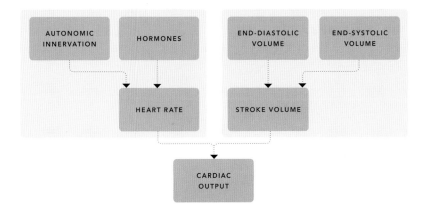

▲ FACTORS CONTROLLING CARDIAC OUTPUT

Several factors act together to influence cardiac output, which is essentially the product of heart rate and stroke volume. Heart rate can be raised by sympathetic neural stimulation (autonomic innervation), or by circulating epinephrine (a cardioregulatory hormone) released from the adrenal medulla. Conversely, heart rate can be lowered by parasympathetic stimulation from the vagus nerve (autonomic innervation). Stroke volume is the amount of blood ejected from the left ventricle with each heart beat. It is the difference in the internal volume of the left ventricle before contraction begins (end-diastolic volume) and the volume after contraction has been finished (end-systolic volume). End-diastolic volume is determined by the rate of venous return and has an effect on the force of cardiac contractility by the Frank-Starling curve. This means that the more venous blood has returned to the heart to dilate the ventricle, the more forcefully the left ventricle will contract with the next beat. End-systolic volume can be affected by the afterload on the heart. If there is high resistance to the ejection of blood from the heart, e.g., because of elevated peripheral resistance or a narrowed aortic valve, then cardiac output may be reduced.

Capillary structure & function

THE CAPILLARIES OF THE BODY ARE SITES WHERE NUTRIENTS, waste products, and gases are exchanged between the bloodstream and the tissues. The structure of capillaries reflects this vital role in exchange: capillary walls are only one cell thick, i.e., simply a single layer of endothelial cells and thin connective tissue. Capillaries also tend to form complex intertwined networks known as capillary beds to ensure that all parts of the tissue being supplied are no more than a few tens of μm (micrometers) away from the bloodstream.

The flow of blood through the capillary beds is referred to as the microcirculation. Blood flowing into the capillary bed comes from arterioles that branch into capillaries, which later rejoin to form venules. The body has the capacity to modify the amount of blood that passes through a capillary bed by the action of tiny precapillary sphincters that surround the arteriolar end of capillaries and can be opened or closed as required. As part of the microcirculation, there are also vascular shunts that directly connect the supplying arterioles with the venules.

When it is not suitable to perfuse the entire capillary bed, e.g., in the skin where perfusion of the dermal capillary bed might lead to excess heat loss in cold weather, the precapillary sphincters are shut and blood is directly shunted from the arteriole to the venule. When it is desirable to pass more blood through the capillary bed, e.g., in hot weather when passing blood through the dermal capillary bed can assist in heat loss, the precapillary sphincters are opened and blood flows through the entire capillary bed.

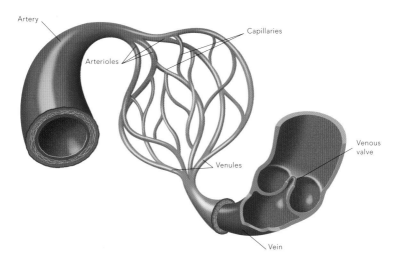

Artery

Arterioles

Capillaries

Venous valve

Venules

Vein

▶ THE MICROCIRCULATION

For blood to supply the tissues of the body, it must pass through a fine network of vessels called the capillary bed. Blood from arteries flows through the arterioles, which have thick smooth muscle walls so they can be shut down as required, and then into the network of capillaries where gas, nutrient, and waste exchange occur. This region of the circulation is called the microcirculation. Capillaries connect to form venules (very small veins), which in turn join to form larger veins. Veins are low-pressure vessels that have internal valves to prevent the backflow of blood, particularly when the blood is being pumped up from the lower limbs.

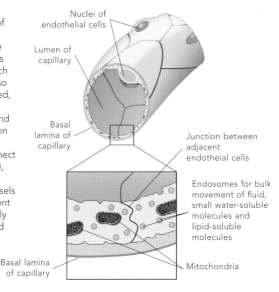

Nuclei of endothelial cells

Lumen of capillary

Basal lamina of capillary

Junction between adjacent endotheial cells

Endosomes for bulk movement of fluid, small water-soluble molecules and lipid-soluble molecules

Basal lamina of capillary

Mitochondria

Starling forces in the capillary bed

GASES SUCH AS OXYGEN AND CARBON DIOXIDE CAN MOVE ACROSS THE CAPILLARY WALL BY SIMPLE DIFFUSION, but there are other mechanisms that facilitate the exchange of nutrients between the blood in capillary beds and the tissues. Many capillaries have tiny windows in their walls (fenestrations) that make them slightly leaky and permit some movement of fluid, dissolved ions, and small molecules across the capillary wall. Large molecules like the plasma protein albumin are too big to pass through the fenestrations and usually stay within the capillary.

Within the capillary bed, there is a balance between the hydrostatic pressure of the fluid in the capillary lumen that tends to push fluid out into the surrounding tissue and the osmotic (also called oncotic) pressure due to the plasma proteins that tends to draw fluid back into the capillary. At the arteriolar end of the capillary bed, the hydrostatic pressure is high (about 35 mm Hg) and the osmotic pressure due to the plasma proteins is lower (about 25 mm Hg), so the net movement of fluid and dissolved substances is out of the capillary. At the venule end of the capillary bed, the hydrostatic pressure is low (only about 17 mm Hg) and the osmotic pressure is about the same as at the arteriolar end (about 25 mm Hg), so fluid tends to be sucked back into the capillary.

The amount of fluid moving out of the arteriolar end of the capillary bed is slightly in excess of that returning at the venule end, so some fluid is left in the extracellular spaces. This fluid passes along the lymphatic channels to eventually return to the bloodstream at the top of the chest cavity.

▶ **THE BALANCE OF FORCES IN THE CAPILLARY BED**
When blood flows through the capillary bed, there is a balance between the movement of fluid out of the vessels due to hydrostatic pressure and the movement of fluid into the vessels due to osmotic (also called oncotic) pressure.

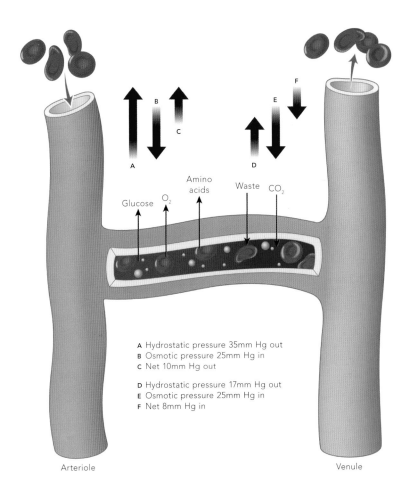

Glucose O_2 Amino acids   Waste CO_2

A Hydrostatic pressure 35mm Hg out
B Osmotic pressure 25mm Hg in
C Net 10mm Hg out

D Hydrostatic pressure 17mm Hg out
E Osmotic pressure 25mm Hg in
F Net 8mm Hg in

Arteriole

Venule

Venous storage & return of blood to the heart

THE ARTERIAL SIDE OF THE SYSTEMIC CIRCULATION IS THE HIGH-PRESSURE SIDE, whereas the venous side of the systemic circulation is adapted for storing fluid in venous channels. The total circulatory blood volume is about 10.5 pints (5 liters), of which about 70% is held within venous channels. Nevertheless, the amount of blood returning from the venous side of the systemic circulation must exactly balance that being pumped out in the arteries, or cardiac output cannot be maintained. If venous blood is allowed to pool in the peripheral circulation rather than return to the heart, the individual will faint and perhaps die.

Pressure within veins is low all the time, so additional mechanisms are required to ensure that venous blood is returned to the heart. Most veins are provided with valves to ensure that venous blood always flows in the direction of the heart. Veins in the lower limb are arranged in superficial and deep networks, with the blood in the superficial channels directed toward the deep ones by the valves. Deep veins in the leg and thigh are also surrounded by muscle, which provides a pumping mechanism to maintain venous return.

An additional pumping effect occurs once venous blood reaches the chest. Drawing air into the lungs lowers the pressure in the thoracic cavity relative to the abdominal cavity, forcing blood from the abdomen into the chest and then into the right side of the heart.

VARICOSE VEINS

Prolonged elevated pressure in the veins of the lower limb can cause these vessels to dilate. If leg veins become too dilated, their valves will cease to be effective and even more blood will accumulate in the lower limb veins, producing tortuous, ropy, and engorged veins known as varicose veins. Elevated pressure in leg veins can also interfere with blood flow through the capillary beds of the lower limb, reducing the ability of the skin of the leg to heal after minor bumps and abrasions. This can lead to venous stasis ulcers on the shins.

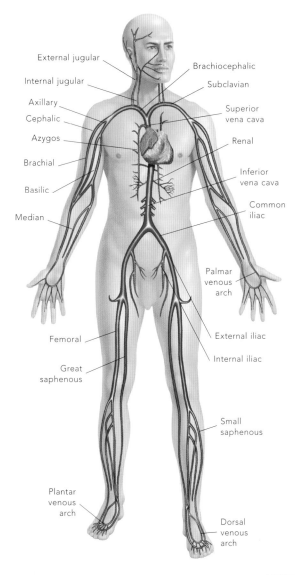

External jugular

Internal jugular

Axillary

Cephalic

Azygos

Brachial

Basilic

Median

Femoral

Great
saphenous

Plantar
venous
arch

Brachiocephalic

Subclavian

Superior
vena cava

Renal

Inferior
vena cava

Common
iliac

Palmar
venous
arch

External iliac

Internal iliac

Small
saphenous

Dorsal
venous
arch

◀ **VEINS OF
THE BODY**

The systemic veins
of the body form
two groups: those
draining blood
from the upper
trunk and upper
limbs into the
superior vena cava
and those draining
blood from the
lower trunk and
lower limbs into
the inferior vena
cava. Most veins
are deep within the
limbs or trunk and
accompany major
arteries, but some
are superficial, e.g.,
the great and small
saphenous veins
of the lower limbs
and the basilic and
cephalic veins of
the upper limb.

Lymphatic system

AT THE ARTERIOLAR END OF THE BED, SLIGHTLY MORE FLUID LEAVES THE CAPILLARIES THAN RETURNS TO THE CAPILLARIES AT THE VENOUS END. This excess fluid must be absorbed from the tissue spaces and eventually returned to the venous side of the circulation. If this were not done, the peripheral tissues would swell with the accumulated fluid in a process called edema.

There are benefits in this excess fluid being sampled because it allows the body's immune system to check for the presence of foreign material and invading microorganisms in the peripheral parts of the body. It is for this reason that lymph nodes containing immune system cells are arranged along the lymphatic channels from the limbs and the internal organs. In the limbs, these lymph nodes are clustered at the bases of the limbs, i.e., in the armpit or axilla, and in the groin or inguinal region, but most lymph nodes are located within the abdominal and thoracic cavities and drain lymph fluid from the gastrointestinal tract and lungs.

Lymph fluid re-enters the venous side of the circulation at the top of the chest cavity, where two major lymph channels (the thoracic duct and the right lymphatic duct) enter the junction of the internal jugular and subclavian veins, where those two veins form the left and right brachiocephalic veins.

▶ **MAJOR LYMPHATIC CHANNELS OF THE BODY**

The major lymphatic channels of the body usually accompany the large veins. Lymph nodes are located along the course of these lymphatic channels so that the immune system can sample the lymph fluid for disease-causing microorganisms and cancer cells.

LYMPHEDEMA

If the excess tissue fluid that leaves the capillary bed is unable to return via the lymphatic channels, swelling of the soft tissues of the limbs (usually lower) will result, leading to a condition called lymphedema. Lymphedema occurs when lymphatic channels are blocked by tumor cells, e.g., when carcinoma of the breast spreads to axillary nodes, or by parasites, e.g., elephantiasis due to the filarial parasite.

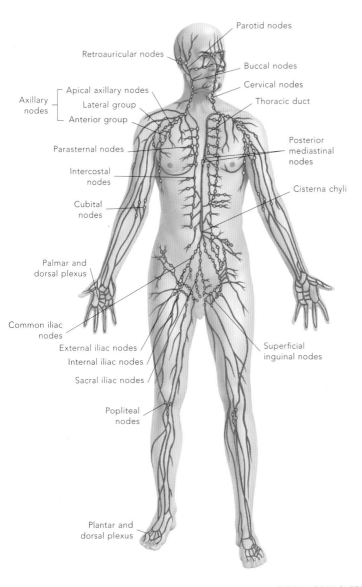

Parotid nodes

Retroauricular nodes

Buccal nodes

Cervical nodes

Thoracic duct

Axillary nodes
- Apical axillary nodes
- Lateral group
- Anterior group

Parasternal nodes

Posterior mediastinal nodes

Intercostal nodes

Cisterna chyli

Cubital nodes

Palmar and dorsal plexus

Common iliac nodes

External iliac nodes

Internal iliac nodes

Sacral iliac nodes

Superficial inguinal nodes

Popliteal nodes

Plantar and dorsal plexus

Effects of hemorrhage & postural change on blood pressure

THE ABILITY TO MAINTAIN EFFECTIVE PERFUSION OF VITAL ORGANS WHEN BLOOD IS LOST FROM THE CIRCULATION IS CRITICAL FOR SURVIVAL. The body can respond to this situation either through the endocrine system or via neural mechanisms.

The endocrine response involves several hormones that either promote sodium and fluid retention, e.g., antidiuretic hormone or ADH—also known as vasopressin, angiotensin II, and aldosterone; act on arterial smooth muscle (angiotensin II); or promote the production of more red blood cells (erythropoietin, or EPO). The net effect of these is to restore blood volume.

The neural response can act through the stimulation of baroreceptors and chemoreceptors (blood pressure and O2 tension in blood, respectively), through direct stimulation of cardioregulatory centers in the brainstem, or through general sympathetic stimulation and release of epinephrine from the adrenal medulla.

The net effect of these is to increase cardiac output and peripheral resistance and to decrease the amount of blood in the venous reserve compartment. The combined action of the hormonal and neural responses restores blood volume and blood pressure to normal. The neural effects are almost immediate, whereas the hormonal effects may take days to weeks to be fully accomplished.

▶ **FLOW DIAGRAM SHOWING THE BODY'S RESPONSE TO BLOOD LOSS**

When blood is lost from the body, several mechanisms are activated to restore blood volume and blood pressure. These fall under the categories of endocrine or humoral responses (on the left) and neural or autonomic responses (on the right). The end result is that blood volume is restored rapidly (within minutes), although the circulating red blood cell volume will take days to weeks to replenish.

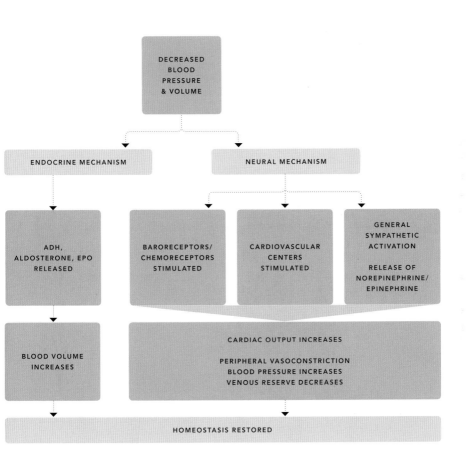

DECREASED
BLOOD
PRESSURE
& VOLUME

ENDOCRINE MECHANISM

NEURAL MECHANISM

ADH,
ALDOSTERONE, EPO
RELEASED

BARORECEPTORS/
CHEMORECEPTORS
STIMULATED

CARDIOVASCULAR
CENTERS
STIMULATED

GENERAL
SYMPATHETIC
ACTIVATION

RELEASE OF
NOREPINEPHRINE/
EPINEPHRINE

BLOOD VOLUME
INCREASES

CARDIAC OUTPUT INCREASES

PERIPHERAL VASOCONSTRICTION
BLOOD PRESSURE INCREASES
VENOUS RESERVE DECREASES

HOMEOSTASIS RESTORED

Structure of the respiratory tract: macroscopic— major airways & respiratory muscles

THE RESPIRATORY SYSTEM CONSISTS OF A CONDUCTING PORTION FOR BRINGING THE GASES OF THE EXTERNAL ENVIRONMENT INTO THE BODY and an exchange or respiratory portion for transfer of gases between the inhaled air and the bloodstream. The conducting portion includes the nasal cavity, nasopharynx, oropharynx, larynx, trachea, left and right bronchi, and lobar bronchi. Beyond the lobar bronchi, there are about 20 additional divisions of the conducting airways, including bronchioles and terminal bronchioles, within the lungs until the exchange part of the respiratory system is reached (see pp. 224–225).

The left lung is divided into two lobes (upper and lower), and the right lung is divided into three (upper, middle, and lower). Both lungs sit within pleural sacs, which are double-layered sacs lined with smooth pleural mesothelium and filled with a thin film of fluid to allow free movement of the lungs within the chest wall. Outside the pleural sacs is the chest wall made of the 12 pairs of ribs, vertebral column, and sternal bones, as well as the intercostal muscles between the ribs. The two lungs sit on the respective domes of the diaphragm and are expanded when the diaphragm domes descend.

Other important respiratory muscles are the intercostal muscles. The outer layer of muscles (external intercostal muscles) elevate the ribs and increase the front-to-back dimension of the chest to produce inspiration, whereas the inner layers (internal and innermost intercostal muscles) lower the ribs and act as expiratory muscles.

▶ **MACROSCOPIC STRUCTURE OF THE RESPIRATORY TRACT**
The respiratory system requires a conducting system to bring air into the lungs, an exchange area in the lungs for transfer of oxygen and carbon dioxide between the inhaled air and the bloodstream, and a musculoskeletal apparatus for ventilation of the lungs.

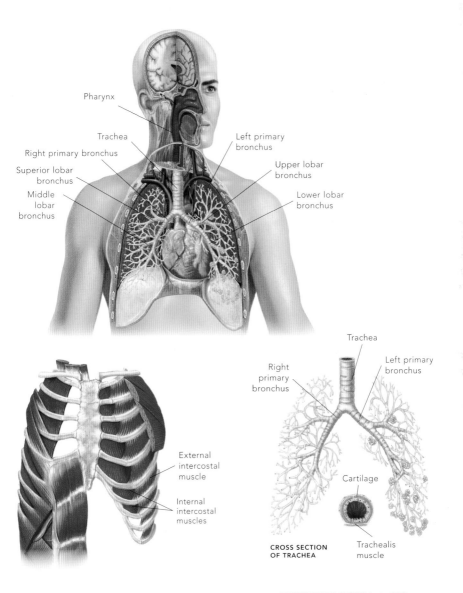

Pharynx

Trachea

Right primary bronchus

Superior lobar bronchus

Middle lobar bronchus

Left primary bronchus

Upper lobar bronchus

Lower lobar bronchus

External intercostal muscle

Internal intercostal muscles

Trachea

Right primary bronchus

Left primary bronchus

Cartilage

Trachealis muscle

CROSS SECTION OF TRACHEA

Structure of respiratory tract: microscopic—alveolar wall

THE EXCHANGE PART OF THE LUNGS CONSISTS PRIMARILY OF THE ALVEOLI OF THE LUNGS, where blood in the capillaries is in close proximity to the gases within the alveolar spaces. When the walls of the respiratory tract are at their thinnest, i.e., in the individual alveoli of the alveolar sacs, there is less than 1 μm (one-thousandth of a mm) between the blood and the air that is in the alveoli.

The wall of the typical alveolus is made up of three types of cells. The type I alveolar cells are flattened or squamous cells that make up about 90% of the cells in the wall. Type II alveolar cells are cuboidal in shape and make the precursor of the chemical surfactant that reduces the surface tension in the alveoli (see pp. 228–229). Alveolar macrophages are phagocytes that are made in the bone marrow. They roam the alveolar spaces, searching out cellular debris, foreign material such as inhaled dust, and invading microorganisms to engulf and remove. When alveolar macrophages have completed their task, they migrate up the airways to be coughed up and swallowed.

The tissue between the inhaled air and the bloodstream is known as the respiratory membrane. It is made up of the very thin type I alveolar cells that line the alveoli, a thin connective tissue layer (the basal lamina), and the endothelial cells that line the capillaries of the lungs. Collectively, these three layers are between 0.2 and 0.6 μm thick so that gases can readily diffuse.

EMPHYSEMA

Emphysema belongs to a group of conditions known as chronic obstructive pulmonary disease (COPD), often caused by smoking. In emphysema, the walls of the alveoli break down, which not only reduces the available surface area for gas exchange but also reduces the lungs' elasticity. The latter leads to hyperinflation of the lungs and a barrel-shaped chest. The loss of surface area for gas exchange causes shortness of breath (dyspnoea).

▶ MICROSCOPIC STRUCTURE OF THE TERMINAL-CONDUCTING AIRWAYS AND ALVEOLI

There are 23 divisions of the airways until the alveoli are reached. Significant gas exchange does not occur until the alveolar sacs and alveoli of the exchange part of the lungs are reached. The exchange portion of the lungs is richly vascularized, receiving branches of the pulmonary artery and being drained by tributaries of the pulmonary veins.

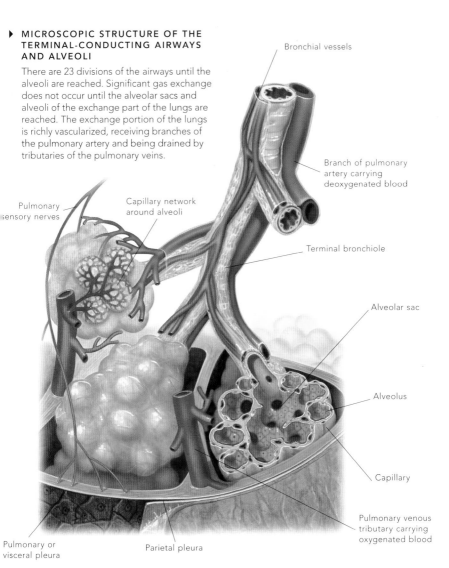

Bronchial vessels

Branch of pulmonary artery carrying deoxygenated blood

Pulmonary sensory nerves

Capillary network around alveoli

Terminal bronchiole

Alveolar sac

Alveolus

Capillary

Pulmonary venous tributary carrying oxygenated blood

Pulmonary or visceral pleura

Parietal pleura

Biomechanics of lung ventilation

THE LUNGS ARE ENCLOSED IN A SMOOTH, DOUBLE-LAYERED PLEURAL SAC WITH A THIN FILM OF FLUID THAT ALLOWS FREE MOVEMENT OF THE LUNGS WITHIN THE CHEST CAVITY. Breathing in (inspiration) is an active process that is achieved mainly by contraction of the muscular diaphragm between the thoracic and abdominal cavities. As the diaphragm contracts, its dome structure descends, compressing the abdominal organs and increasing the vertical dimension of the chest. This increase in the volume of the chest lowers the pressure inside, expands the lungs, and draws air into the airways. In forced inspiration, there may also be movement of the ribs, which can be drawn upward by the external intercostal muscles.

Expiration is predominantly a passive process because inspiration tends to build up elastic potential energy, e.g., in the compression of the abdominal organs due to diaphragm contraction and the stretching of chest wall structures produced by the inspiratory movements of the ribs. When inspiration ceases, these stretched or compressed structures tend to return to their natural state, so the dimensions of the chest return to normal and air is expelled from the lungs.

When forced ventilation is required, e.g., in exercise or when the lungs are diseased, additional accessory muscles may come into play. These include the pectoralis major and minor and serratus anterior muscles for inspiration, and the anterolateral abdominal wall muscles for expiration.

▶ **MUSCLES INVOLVED IN LUNG VENTILATION**

The major inspiratory muscle is the diaphragm, which is a musculotendinous structure that separates the thoracic and abdominal cavities. Contraction of the muscular parts of the diaphragm make the two domes descend, thereby increasing the vertical dimension of the chest. Expiration is largely a passive process, in which elastic potential energy built up by inspiration is used to reduce the dimensions of the chest and return the lung volume to rest.

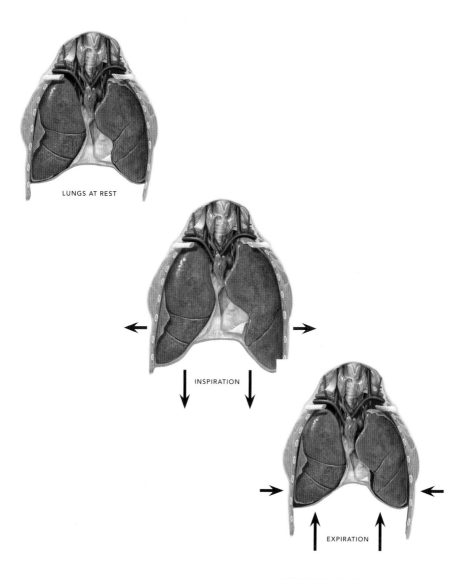

LUNGS AT REST

INSPIRATION

EXPIRATION

Pulmonary surfactants, surface tension, & lung compliance

THE LUNGS HAVE AN ASTONISHINGLY LARGE INTERNAL SURFACE AREA (60 to 84 square yards/50 to 70 m²). The presence of a water film on this surface would generate a substantial surface tension that would tend to collapse the alveolar spaces without the presence of some chemical agent to lower the surface tension. This detergent-like agent is called pulmonary surfactant and is made by the cuboidal type II alveolar cells of the alveolar spaces.

Pulmonary surfactant is not usually produced in the developing lung until about 28 to 30 weeks' gestation, so infants born before that age must expend large amounts of energy to keep their lungs inflated. As a result, they develop Infant Respiratory Distress Syndrome (IRDS).

The elastic recoil of lung tissue is vitally important for the return of lung volumes to normal at the end of each inspiration. In emphysema, the elastic connective tissues of the lungs that produce the elastic recoil are damaged, with the result that the lungs become

▶ **THE ROLE OF PULMONARY SURFACTANT**

The alveolar sacs and alveoli present a huge surface area (one-third of a tennis court) with considerable surface tension. Pulmonary surfactant made by type II alveolar cells reduces the surface tension of the exchange parts of the lungs so that the lungs can be easily inflated.

hyperinflated. Pulmonary compliance is the ability of the lungs and chest wall to stretch. Pulmonary compliance depends on the strength of surface tension, the distensibility of elastic tissue in the lungs, and the ability of the chest wall to move. Factors that reduce pulmonary compliance include diseases that reduce surfactant production, e.g., IRDS. Other diseases, such as tuberculosis and dust diseases, like anthracosis, asbestosis, or silicosis, make the lungs fibrous, stiffer, and more difficult to inflate.

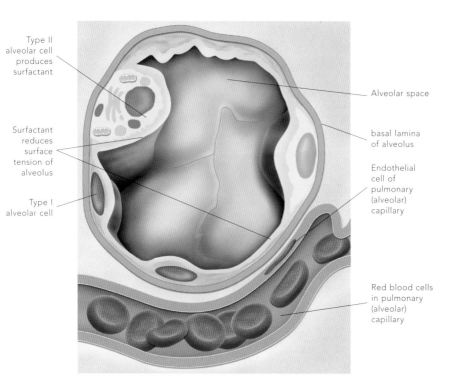

Type II alveolar cell produces surfactant

Surfactant reduces surface tension of alveolus

Type I alveolar cell

Alveolar space

basal lamina of alveolus

Endothelial cell of pulmonary (alveolar) capillary

Red blood cells in pulmonary (alveolar) capillary

INFANT RESPIRATORY DISTRESS SYNDROME

Infant Respiratory Distress Syndrome (IRDS) is a condition in which prematurely born infants have insufficient pulmonary surfactant to inflate their lungs adequately. IRDS affects 1% of newborn infants and is a leading cause of death among premature infants. About half of all babies born before 28 weeks' gestation will be affected. Preventative measures can be taken by giving the mother corticosteroids before delivery, which speeds up surfactant production. Treatment after delivery mainly involves continuous positive airway pressure (CPAP) to assist breathing.

Lung volumes & ventilation

MEASUREMENT OF THE VOLUMES OF AIR THAT A PERSON CAN MOVE INTO AND OUT OF THE LUNGS GIVES IMPORTANT CLUES TO LUNG FUNCTION AND THE PRESENCE OF DISEASE. This information is obtained by a clinical investigation called spirometry. The tidal volume (V_T) is the amount of air (about 1.1 pints/500 ml) that a person breathes in and out during a normal breath. Most adults breathe 12 times per minute so the minute volume (the amount of air breathed in and out every minute) is 13 pints (6 liters).

It should be noted that not all of the air that is breathed in will come into contact with alveoli because some air will remain in the conducting part of the respiratory tract, i.e., the larynx, trachea, and major bronchi. This part of the tidal volume that never comes into contact with the exchange surfaces is called the anatomical dead space and is about 0.3 pints (150 ml). So only 0.7 pints (350 ml) per breath actually comes into contact with the alveoli, and the minute volume of this is called the alveolar ventilation rate (0.7 pints [350 ml] x 12 = 9 pints [4.2 liters]).

Inspiratory reserve volume (IRV) is the volume of air that can be inspired by force beyond the normal tidal inspiration—between 4.4 and 7 pints (2,100 and 3,300 ml). The expiratory reserve volume (ERV) is the amount of air that can be forced out of the lungs after a normal tidal expiration—between 1.5 and 2.5 pints (700 and 1,200 ml). Some air will be left in the lungs after a forced expiration. This volume (about 2.5 pints/1,200 ml) is the residual volume. The full range of air that can be forced into and out of the lungs (vital capacity, or VC) is the sum of ERV + TV + IRV and is 6.5 to 10.2 pints (3,100 to 4,800 ml).

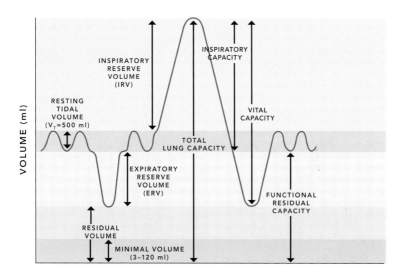

VOLUME (ml)

INSPIRATORY RESERVE VOLUME (IRV)

INSPIRATORY CAPACITY

RESTING TIDAL VOLUME ($V_T = 500$ ml)

VITAL CAPACITY

TOTAL LUNG CAPACITY

EXPIRATORY RESERVE VOLUME (ERV)

FUNCTIONAL RESIDUAL CAPACITY

RESIDUAL VOLUME

MINIMAL VOLUME (3–120 ml)

▲ LUNG VOLUMES AND
VENTILATION VALUES

Lung ventilation at rest exchanges
only 500 ml with the external
environment per breath (resting tidal
volume), but the lungs of a large person
have the capacity to draw in as much as
3.3 liters of inspiratory reserve volume
and can expel as much as 1.2 liters of
expiratory reserve volume.

Regulation of lung ventilation

THE RATE AND DEPTH OF LUNG VENTILATION NEEDS TO BE CLOSELY CONTROLLED TO ENSURE THAT KEY BLOOD GASES—oxygen (O_2) and carbon dioxide (CO_2)—stay within optimal ranges. This control depends on two negative feedback loops—one that comes into play when lung ventilation is insufficient and the other when ventilation is excessive.

When lung ventilation is low (hypoventilation), O_2 drops in the arterial blood, whereas CO_2 builds up, raising the hydrogen ion (H^+) concentration and lowering the pH. Chemoreceptors in the carotid body (for O_2 tension) and the brainstem surface (for CO_2) detect these changes and signal the respiratory control centers in the brainstem. These drive faster and deeper ventilation to blow off CO_2 and increase arterial O_2 tension, feeding back to the original blood gases.

In the converse situation, i.e., when there is hyperventilation, O_2 increases in the arterial blood and CO_2 drops, lowering the hydrogen ion concentration in the blood and raising the blood pH. These changes are detected by the chemoreceptors in the carotid body and brainstem and signal the respiratory control centers in the brainstem to reduce the ventilation rate and depth to return the blood gases to optimal levels.

▶ **FEEDBACK LOOPS THAT MAINTAIN BLOOD GASES AT OPTIMAL LEVELS**

Blood gases (O_2 and CO_2) and hydrogen ion concentration (affecting pH) are maintained at optimal levels in the blood by feedback loops that regulate lung ventilation rate and depth in response to gas concentration. The chemoreceptors for O_2 are located in the carotid bodies at the branching of the common carotid arteries into the external and internal carotid arteries. The chemoreceptors for CO_2 are located in the surface of the medulla oblongata.

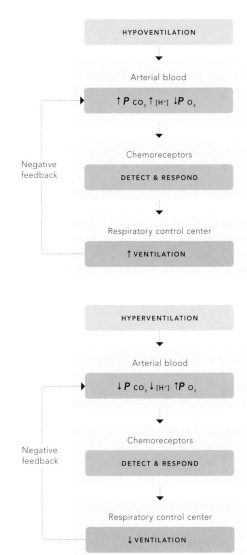

HYPOVENTILATION

Arterial blood

$\uparrow P_{CO_2}$ $\uparrow_{[H^+]}$ $\downarrow P_{O_2}$

Chemoreceptors

DETECT & RESPOND

Respiratory control center

$\uparrow$ VENTILATION

Negative feedback

HYPERVENTILATION

Arterial blood

$\downarrow P_{CO_2}$ $\downarrow_{[H^+]}$ $\uparrow P_{O_2}$

Chemoreceptors

DETECT & RESPOND

Respiratory control center

$\downarrow$ VENTILATION

Negative feedback

P PARTIAL PRESSURE

INITIAL STIMULUS

PHYSIOLOGICAL RESPONSE

RESULT

Diffusion of gases across the alveolar wall

PULMONARY GAS EXCHANGE INVOLVES THE TRANSFER OF GASES BETWEEN THE ALVEOLAR SPACES AND THE HEMOGLOBIN OF THE BLOOD. This process turns the oxygen-poor and carbon dioxide-rich blood that is delivered to the lungs by the blood in the pulmonary artery into oxygen-rich and carbon dioxide-reduced blood that flows back to the left side of the heart by the pulmonary veins.

The transfer of gases is driven by differences in the partial pressure of the respective gases in the alveolar spaces and the pulmonary capillaries. The partial pressure of oxygen in the alveolar spaces is 104 mm Hg, whereas the pressure is only 40 mm Hg in the pulmonary capillaries. The partial pressure of carbon dioxide in the alveolar spaces is 40 mm Hg, whereas the partial pressure of carbon dioxide in the pulmonary capillaries is 45 mm Hg.

Gas pressure gradients are not the only determinants of gas exchange at the whole lung level. Other key factors are the available surface area of the respiratory membrane (60 to 80 square yards/50 to 70 m²), the thickness of the respiratory membrane (0.2 to 0.6 μm), and the match between the amount of air entering the lungs and the amount of blood flowing through the lungs (ventilation-perfusion matching). Ventilation-perfusion matching is optimized by the direction of blood flow to those parts of the lungs with the best ventilation. Conversely, ventilation is directed to those parts of the lungs receiving the most blood flow.

ASTHMA

Asthma is a condition of airway constriction and submucosal swelling that is usually due to hyperactivity of smooth muscle and glands in the airway wall. Symptoms include wheezing, shortness of breath, and tightness of the chest. Exposure to allergens, such as pollens, or to exercise, some chemicals, and cold air are common causes. Treatment is by drugs that stabilize or depress the immune response and sympathomimetic agents that dilate the airways. Corticosteroids may be used in severe cases.

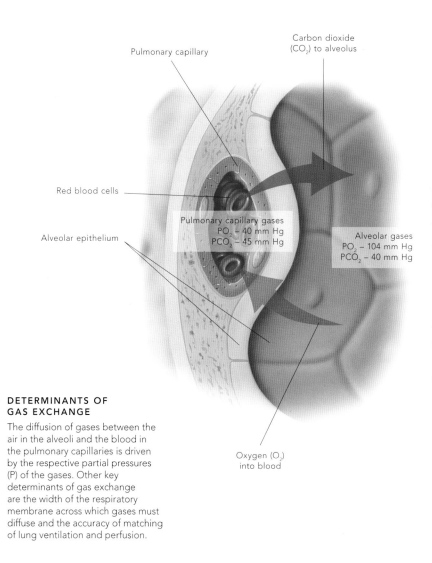

Pulmonary capillary

Carbon dioxide
(CO_2) to alveolus

Red blood cells

Alveolar epithelium

Pulmonary capillary gases
PO_2 – 40 mm Hg
PCO_2 – 45 mm Hg

Alveolar gases
PO_2 – 104 mm Hg
PCO_2 – 40 mm Hg

Oxygen (O_2)
into blood

▶ DETERMINANTS OF GAS EXCHANGE

The diffusion of gases between the air in the alveoli and the blood in the pulmonary capillaries is driven by the respective partial pressures (P) of the gases. Other key determinants of gas exchange are the width of the respiratory membrane across which gases must diffuse and the accuracy of matching of lung ventilation and perfusion.

Hemoglobin: structure & function

HEMOGLOBIN (HB) IS AN OXYGEN-CARRYING PROTEIN IN RED BLOOD CELLS THAT CONSISTS OF FOUR PROTEIN (GLOBIN) CHAINS, each containing a heme group. In adult hemoglobin, the protein chains are two alpha chains and two beta chains. The heme group is a porphyrin molecule with a single iron atom inside its organic ring, and this iron atom can bind to just one molecule of oxygen. The binding and release of oxygen to Hb are known as loading and unloading reactions, respectively. The unloaded form of Hb is known as deoxyhemoglobin, and the loaded form is called oxyhemoglobin.

The affinity or binding tendency of Hb for oxygen changes with the oxygen partial pressure. When Hb is fully saturated with oxygen, it tends to hang onto its oxygen molecules tightly. As oxygen tension drops, e.g., when the oxygenated blood reaches peripheral capillaries, and Hb loses its first molecule of oxygen, the change in the shape of the four globin chains makes it easier for the second and third molecules of oxygen also to be given up. This means that when Hb is flowing through parts of the body where oxygen is low, the Hb will give up its oxygen readily. Nevertheless, the very last oxygen molecule is only given up when oxygen levels in the vicinity of the Hb molecule are very low.

This tendency of the Hb molecule to change its oxygen affinity with surrounding oxygen concentration makes the shape of the relationship between percentage oxygen saturation of Hb and the partial pressure of oxygen a sigmoid, or "S"-shaped, curve (see fetal hemoglobin pp. 238–239).

▶ **THE STRUCTURE OF HEMOGLOBIN**

Hemoglobin in adults consists of four protein (globin) chains (two alpha and two beta chains), each with a heme porphyrin molecule in its center. Each heme molecule can bind a single molecule of oxygen. The shape of the globin molecules can change with ambient oxygen tension, adjusting the ease of oxygen release to the local requirements.

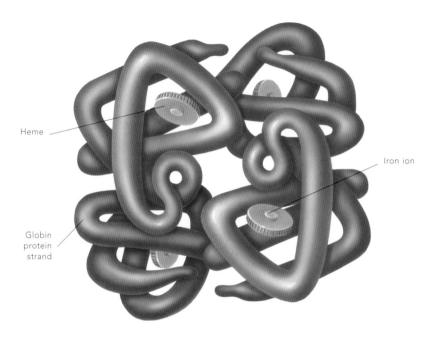

Heme

Iron ion

Globin
protein
strand

ANEMIA

Anemia is a condition characterized by insufficient red cell mass and therefore a
limited capacity to carry oxygen in the blood. Anemia can be due to problems
with production of hemoglobin and red blood cells, e.g., iron deficiency anemia
and vitamin B12 deficiency; faulty hemoglobin, e.g., hemoglobinopathies like
thalassemia; or excess destruction or loss of red blood cells, e.g., enlarged spleen
and blood loss. Many types of anemia exhibit a characteristic red cell appearance.
These include the small, lightly colored red blood cells of iron deficiency anemia
(hypochromic microcytic anemia) and the large immature red blood cells of
vitamin B12 deficiency (megaloblastic cells of pernicious anemia).

Fetal hemoglobin

ADULT HEMOGLOBIN IS OPTIMIZED FOR THE TRANSPORT OF OXYGEN AT THE CONCENTRATIONS OF OXYGEN FOUND IN THE NORMAL ADULT BLOODSTREAM, but oxygen concentration in the blood during fetal life is very different. Fetal hemoglobin (HbF) must function at very low oxygen tensions (typically 70% of that in the adult circulation) because oxygen comes from gas transfer with the maternal circulation across the placenta. For this reason, HbF is adapted to take on oxygen and release it at much lower oxygen partial pressures than adult Hb. HbF is 50% saturated with oxygen at about 19 mm Hg, whereas adult Hb is 50% saturated at about 27 mm Hg oxygen tension.

The greater affinity of HbF for oxygen is due to its different structure. In HbF, the beta globin chains are replaced with gamma chains so their Hb consists of two alpha chains and two gamma chains. The site of production of Hb and red blood cells is also different before birth. Prenatally, red blood cell production is initially within the embryonic yolk sac (first three to four months of development) and later within the fetal liver (last five months of prenatal life). Production of red blood cells in the bone marrow plays a major role only from birth onward. HbF is completely replaced by adult Hb by about six months of age.

▶ **PRODUCTION AND FUNCTION OF FETAL HEMOGLOBIN**

Fetal hemoglobin is adapted to release oxygen at lower partial pressures than adult hemoglobin. This means that the oxygen dissociation curve for fetal hemoglobin (blue line) is shifted to the left of that for adult hemoglobin (red line). During development, red blood cell production begins in the yolk sac of the embryo before shifting to the fetal liver and spleen. Blood cell production in the bone marrow ramps up in the last trimester. Embryonic and fetal hemoglobin use the beta and gamma globin chains, which are usually only made during the first trimester and first year after conception, respectively.

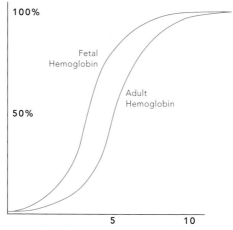

% SATURATION OF HEMOGLOBIN WITH OXYGEN

100%

Fetal Hemoglobin

Adult Hemoglobin

50%

5   10

PARTIAL PRESSURE OF OXYGEN (KPa)

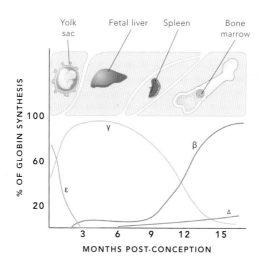

Yolk sac   Fetal liver   Spleen   Bone marrow

% OF GLOBIN SYNTHESIS

100

γ

β

60

ε

20

Δ

3   6   9   12   15

MONTHS POST-CONCEPTION

Oxygen transport by the blood

ABOUT 280 MILLION HEMOGLOBIN (HB) MOLECULES ARE PRESENT IN EACH RED BLOOD CELL, and when the Hb concentration of the blood is about 15 g/dl (15 g/100 ml), the oxygen carried in the blood can be up to 20.4 ml per 100 ml of blood. Most oxygen is transported in the blood combined with hemoglobin as oxyhemoglobin, but a small amount is dissolved in the blood plasma, outside the red blood cells.

Impaired oxygen transport to the body tissues is called hypoxia. When oxygen tension in the capillary beds at the tissue level is low, the mucosa of the oral cavity and tongue and the tissues of the nail beds take on a bluish tinge (cyanosis).

Carbon monoxide is a colorless, odorless gas that is extremely dangerous and is a leading cause of death from fire. Poisoning with carbon monoxide can greatly impair oxygen transport because the carbon monoxide molecule combines with the iron atom inside the heme group with even greater affinity (about 200 to 230 times) than the oxygen molecule.

Carbon monoxide poisoning does not produce cyanosis or respiratory distress; the poisoned individual experiences confusion and headache before succumbing to seizures, coma, and death.

▶ **METHODS OF OXYGEN TRANSPORT**

Most oxygen carried in the blood is transported bound to hemoglobin as oxyhemoglobin. Only about 3% of the dissolved oxygen is transported in the dissolved state. However, the fraction of oxygen bound to hemoglobin that is given up in tissue capillaries when the body is at rest is only about 25%, although this utilization fraction can rise to 75% in strenuous exercise.

OXYGEN TRANSPORT BY BLOOD

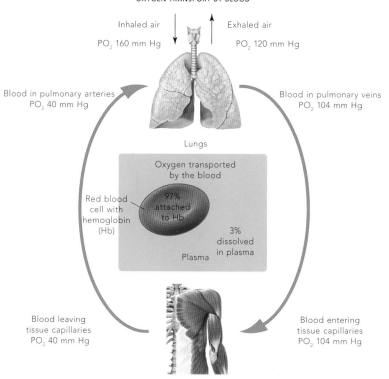

Inhaled air
PO_2 160 mm Hg

Exhaled air
PO_2 120 mm Hg

Blood in pulmonary arteries
PO_2 40 mm Hg

Blood in pulmonary veins
PO_2 104 mm Hg

Lungs

Oxygen transported
by the blood

Red blood
cell with
hemoglobin
(Hb)

97%
attached
to Hb

3%
dissolved
in plasma

Plasma

Blood leaving
tissue capillaries
PO_2 40 mm Hg

Blood entering
tissue capillaries
PO_2 104 mm Hg

Tissue
e.g., skeletal muscle

Transport of carbon dioxide & hydrogen ions by the blood

ANOTHER IMPORTANT FUNCTION OF THE BLOOD AND HEMOGLOBIN (HB) IS TO TRANSPORT CARBON DIOXIDE. This is produced by oxidative metabolism in the mitochondria of the tissues of the body as they make ATP for the many different cellular functions.

Some CO_2 (about 10%) is dissolved in the blood plasma, but most (about 70%) is transported in the plasma as the bicarbonate ion (HCO_3^-). This ion is also very important in buffering the blood to maintain optimal pH (between 7.35 and 7.45). Bicarbonate ions are actually formed within red blood cells by the action of an enzyme called carbonic anhydrase. The newly formed bicarbonate ions are then transported to the blood plasma outside the red blood cells by an HCO_3^-/Cl^- exchange pump.

About 20% of the CO_2 transported in the blood is bound with Hb in the red blood cells, but at a different site on the Hb molecule than oxygen. This means that CO_2 transport doesn't interfere with oxygen transport. When carbon dioxide needs to be released at the lungs, HCO_3^- must first re-enter the red blood cells and combine with hydrogen ions (H^+) to be converted back to carbon dioxide and water. The carbon dioxide then diffuses across the red blood cell membrane and across the alveolar wall to the alveolar space.

▶ **METHODS OF CO_2 TRANSPORT**

Transfer of carbon dioxide between the tissue and blood and between the blood and the alveoli is driven by differences in partial pressure between the different compartments. Carbon dioxide can be transported in the blood in several different ways. About 20% is bound to hemoglobin as carbaminohemoglobin. Most (about 70%) is converted to bicarbonate ions by carbonic anhydrase in the red blood cells, and the remainder is carried dissolved in the plasma.

TRANSFER OF CO₂ FROM PULMONARY CAPILLARY TO ALVEOLUS

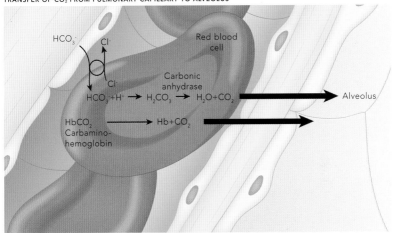

TRANSFER OF CO₂ FROM TISSUES TO BLOOD

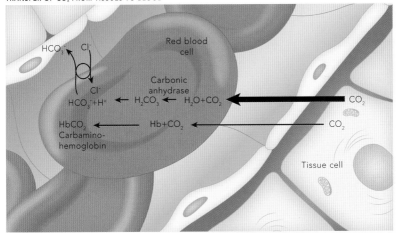

Overview of the digestive system

THE DIGESTIVE SYSTEM IS ESSENTIALLY A TUBULAR STRUCTURE WITH ASSOCIATED GLANDS ATTACHED AS SIDE ORGANS, e.g., salivary glands, liver, and pancreas. Anatomically and functionally, the digestive system consists of components concerned with the ingestion of food (teeth, tongue, and oral cavity), the movement of food from the mouth to the stomach (pharynx and esophagus), and the digestion of food (salivary glands, stomach, exocrine pancreas, liver, gall bladder, and bile ducts for bile storage and release, and enzymes produced by the small intestine—duodenum, jejunum, and ileum). It is also involved in the absorption of food and water mainly within the small intestine, but also within the cecum and various parts of the colon for absorption of water and minerals, and the excretion of waste products principally via the rectum and anus.

The wall of the gut tube has a similar structure throughout its length with some modifications. The common features of gut structure are the presence of a lining layer (the mucosa),

▶ STRUCTURE OF THE DIGESTIVE SYSTEM

The digestive system includes an upper component (teeth, tongue, salivary glands, pharynx, and esophagus), but most of the system is located within the abdominal cavity. The upper abdomen contains the stomach, liver, pancreas, and gall bladder. The lower abdomen is dominated by the intestines, with the large intestine surrounding the small intestine like a picture frame.

sometimes with finger-like projections (villi) and folds (plicae circulares) specialized for digestion and absorption; a submucosa that usually contains glands to produce enzymes for digestion; a double or triple layer of smooth muscle (usually an inner circular layer and an outer longitudinally running layer) to move the gut contents along the tube; and an outer serosal layer that is often suspended by a mesentery from the body wall.

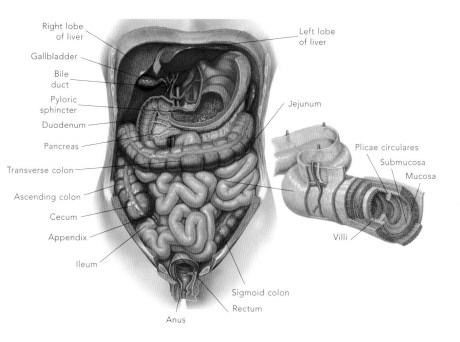

Right lobe of liver
Left lobe of liver
Gallbladder
Bile duct
Pyloric sphincter
Jejunum
Duodenum
Pancreas
Plicae circulares
Submucosa
Mucosa
Transverse colon
Ascending colon
Cecum
Appendix
Villi
Ileum
Sigmoid colon
Rectum
Anus

CROHN'S DISEASE

Crohn's disease is an inflammatory bowel disease that can affect any part of the gut tube between the oral cavity and the anus and usually starts in the teens and twenties. The symptoms include abdominal pain, diarrhea, rashes, and fever. Crohn's disease occurs in genetically susceptible people (about half the risk) who are exposed to environmental, bacterial, and immune system factors. Abscesses, bowel obstruction, and cancers may occur as complications. Treatment is by anti-inflammatory agents like corticosteroids or surgery where abscesses or obstruction occur. Recent treatments using antibody-based agents offer significant improvements in managing the condition.

Salivary gland structure & function

SALIVA CONSISTS OF WATERY FLUID WITH DISSOLVED ENZYMES AND MUCUS. Saliva serves many roles: moistening food so it can be swallowed, allowing digestive enzymes to begin the breakdown of starch, providing a fluid phase for tastants to reach taste buds, and providing a fluid seal when we suck on straws.

Saliva is produced in minor and major salivary glands. The former are visible in the mouth wall only under the microscope, but the latter are large glands that have ducts draining into the oral cavity. The parotid gland is located immediately in front of the ear and has a duct (the parotid duct) that drains forward to the oral cavity to open opposite the crown of the second upper molar. The submandibular and sublingual glands are located beneath the mandible and the tongue, respectively. The submandibular gland has a single duct that opens into the floor of the mouth under the tongue, and the sublingual gland has multiple ducts opening into the oral cavity floor directly above the gland. Saliva contains the enzyme amylase that

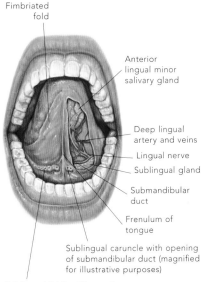

Fimbriated fold

Anterior lingual minor salivary gland

Deep lingual artery and veins

Lingual nerve

Sublingual gland

Submandibular duct

Frenulum of tongue

Sublingual caruncle with opening of submandibular duct (magnified for illustrative purposes)

Sublingual folds with openings of sublingual ducts (magnified for illustrative purposes)

▶ **ORAL CAVITY AND SALIVARY GLANDS**

The major salivary glands are located in front of the ear (parotid gland) or under the tongue (submandibular and sublingual glands). The composition of the salivary fluid produced by each differs, and this is reflected in the glandular structure.

begins the process of breaking down starch into smaller molecules. It also contains the enzyme lysozyme that is capable of perforating bacterial membranes and helps other bactericidal compounds to penetrate bacterial cells.

Secretory immunoglobulin A (IgA) in saliva binds to antigens on harmful bacteria and begins their destruction. Saliva also contains bicarbonate ions that help neutralize any stomach acid that regurgitates into the esophagus.

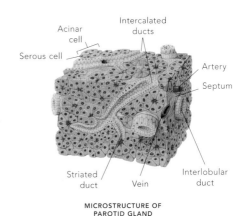

MICROSTRUCTURE OF PAROTID GLAND

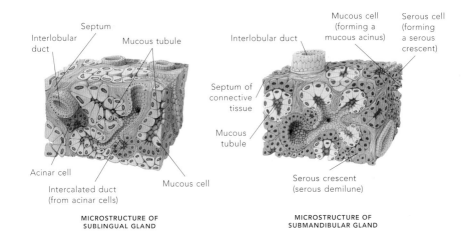

MICROSTRUCTURE OF SUBLINGUAL GLAND

MICROSTRUCTURE OF SUBMANDIBULAR GLAND

Pharynx & esophagus: physiology of swallowing

ONCE FOOD HAS BEEN BROKEN UP BY THE TEETH AND MIXED WITH THE FLUID AND ENZYMES OF SALIVA, it must be transferred to the stomach for further digestion. Swallowing requires that the food has been formed into a moist bolus by chewing and mixing with salivary fluid. This bolus is then compressed between the back of the tongue and the roof of the mouth (hard and soft palate) and squeezed backward into the pharynx.

The entrance to the airway (laryngeal entrance) must be protected from entry of food or fluid, and this is achieved by a combination of the elevation of the larynx and the closure of the epiglottis over the laryngeal opening. Once the airway has been protected, the muscles of the pharynx squeeze the bolus downward past the airway entrance toward the beginning of the esophagus in the middle of the neck. The esophagus has an arrangement of inner circular and outer longitudinal smooth muscle that gently pushes the food down toward the stomach in a process called peristalsis.

There are several organs in contact with the outside of the esophagus that cause compressions and may contribute to difficulty in swallowing (dysphagia). These are at the beginning of the esophagus, in the middle of the chest at the arch of the aorta, and where the esophagus passes through the diaphragm.

▶ **STEPS IN THE PROCESS OF SWALLOWING**

Swallowing requires that the food has been formed into a moist bolus. This is then squeezed between the palate and back of the tongue (upper illustration) to force the bolus into the oropharynx. By this stage, the laryngeal entrance has been closed by elevation of the larynx and bending of the epiglottis so the bolus can slip into the laryngopharynx (middle illustration). Once the bolus has entered the esophagus (lower illustration), peristalsis of the esophageal skeletal and smooth muscle takes over, moving the bolus down to the stomach.

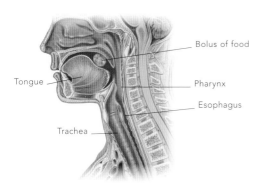

Bolus of food

Tongue

Pharynx

Esophagus

Trachea

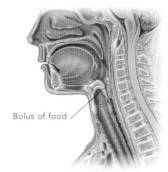

Bolus of food

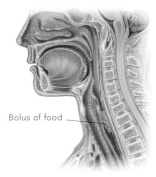

Bolus of food

Structure & function of the stomach

THE STOMACH IS ESSENTIALLY A MUSCULAR SAC THAT CHURNS THE FOOD (mechanical digestion); denatures proteins and foreign material, e.g., viruses and bacteria by hydrochloric acid (chemical digestion); and begins the process of enzymatic digestion of proteins (enzymatic digestion). The stomach receives food from the esophagus and empties into the first part of the duodenum.

The stomach has a thick wall made up of three layers of smooth muscle with fibers arranged in circular, oblique, and longitudinal directions.

The inner lining of the stomach consists of a mucosa with specialized parietal or oxyntic cells that make hydrochloric acid, zymogenic or chief cells that make pepsinogen (the precursor of the protein-digesting enzyme pepsin), and mucus-secreting cells that produce the protective layer of mucus that lines the stomach interior and prevents digestion of the stomach wall by its own secretions.

Beneath the mucosa is the muscularis mucosae—a layer of smooth muscle that moves the mucosa

▶ **MACROSCOPIC STRUCTURE OF THE STOMACH**

The stomach is a sac designed to churn food mixed with acidic secretions. This is reflected in the thick muscular wall of the stomach and the presence of specialized acid-secreting cells (the parietal or oxyntic cells). Once this process is complete, the food is ejected from the stomach through the pyloric sphincter.

into contact with stomach contents to assist digestion. Beneath the muscularis mucosae lies the submucosa with glands that contribute secretions to stomach juices. The outermost layer of the stomach is the serosal layer, which has attachments by peritoneal folds (the greater and lesser omentum) to adjacent organs, i.e., the spleen and transverse colon are attached to parts of the greater omentum, and the liver is attached to the lesser omentum.

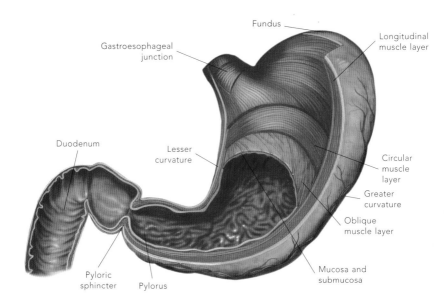

Labels on figure:
Fundus

Gastroesophageal junction

Longitudinal muscle layer

Duodenum

Lesser curvature

Circular muscle layer

Greater curvature

Oblique muscle layer

Pyloric sphincter

Pylorus

Mucosa and submucosa

ULCER

An ulcer is a circumscribed break in an epithelial surface. Peptic ulcers are those that occur where there is contact with gastric juices, usually in the stomach or first part of the duodenum (initial part of the small intestine). Ulcers are a particular hazard in the stomach and the duodenum because of the highly damaging nature of the stomach acid. The stomach lining is protected from autodigestion by a layer of mucus, and the duodenum is protected by the alkaline secretions of the pancreas. Ulcers can cause serious blood loss if a major artery is involved in the ulcer bed and may even cause perforation of the gut wall, leading to chemical and bacterial peritonitis. Infection with *Helicobacter pylori* is a major causative factor in the development of peptic ulcers.

Neural control of gut function

THE GASTROINTESTINAL TRACT IS CONCERNED WITH THE DIGESTION AND ABSORPTION OF NUTRIENTS, so it is not surprising that most of its functions are controlled by that part of the autonomic nervous system concerned with restoring the body to balance— the parasympathetic nervous system.

The vagus nerve from the medulla oblongata is the major nerve controlling gut function, with connection to the gut tube from the level of the soft palate down to the left colic flexure of the large intestine. The vagus nerve increases the secretions from the glands of the gut and stimulates smooth muscle activity to assist in peristalsis—the movement of gut contents from the mouth to the anus. It also controls the pyloric sphincter at the outflow point of the stomach.

The lower part of the gut tube, from the left colic flexure to the anus, is controlled by the parasympathetic outflow from the sacral segments of the spinal cord (S2 to S4). These pelvic nerves control the rectum and anus and play important roles in the control of defecation. There is also a sympathetic nervous system influence on gut function, but its main role is to shut down blood supply to the gut when blood is required for the muscles during emergency situations.

Note that both the parasympathetic and sympathetic nervous systems are not micromanagers because the gut contains its own enteric nervous system (ENS) that can regulate the details of gut function. The parasympathetic and sympathetic nerves are acting more as senior management, giving general instructions on gut function.

▶ **NERVES INVOLVED IN GUT FUNCTION**

The major neural influence on the gut is the parasympathetic vagus nerve that controls gut activity from the esophagus to the left colic flexure. The pelvic splanchnic nerves from sacral spinal cord segments (S2 to S4) control the descending colon, sigmoid colon, and rectum. Sympathetic connections control gut vasculature, but their nerves also carry important sensory information back to the spinal cord for referred pain.

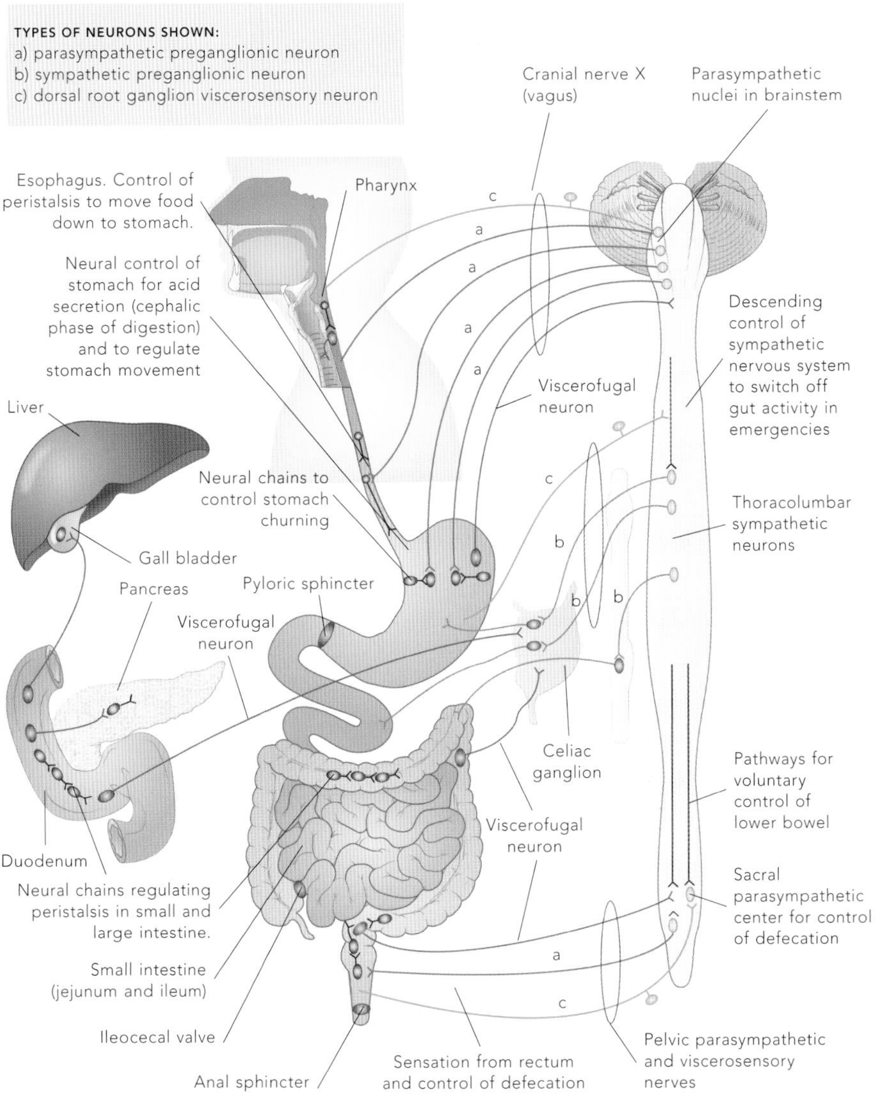

TYPES OF NEURONS SHOWN:
a) parasympathetic preganglionic neuron
b) sympathetic preganglionic neuron
c) dorsal root ganglion viscerosensory neuron

Cranial nerve X (vagus)

Parasympathetic nuclei in brainstem

Esophagus. Control of peristalsis to move food down to stomach.

Pharynx

Neural control of stomach for acid secretion (cephalic phase of digestion) and to regulate stomach movement

Descending control of sympathetic nervous system to switch off gut activity in emergencies

Liver

Viscerofugal neuron

Thoracolumbar sympathetic neurons

Neural chains to control stomach churning

Gall bladder

Pancreas

Pyloric sphincter

Viscerofugal neuron

Celiac ganglion

Duodenum

Viscerofugal neuron

Pathways for voluntary control of lower bowel

Neural chains regulating peristalsis in small and large intestine.

Sacral parasympathetic center for control of defecation

Small intestine (jejunum and ileum)

Ileocecal valve

Anal sphincter

Sensation from rectum and control of defecation

Pelvic parasympathetic and viscerosensory nerves

Hormonal control of gut function

IN ADDITION TO THE NEURAL CONTROL OF GUT FUNCTION, there are a range of circulating hormones that can affect the secretion and motility of the gut. These diffusible agents are mostly small chain proteins or polypeptides and are produced by the inner lining of the gut wall. Some of these are secreted and act locally, whereas others like growth hormone, insulin, and glucagon have widespread effects throughout the body.

Some factors like ghrelin, pancreatic polypeptide, GLP-1, oxyntomodulin, and PYY3-36 play key roles in appetite control and have effects on the central nervous system hunger and satiety centers. Gastrin, secreted by the stomach mucosa, helps regulate acid secretion. Cholecystokinin and secretin, released from the duodenum in response to the presence of ingested food, exert their main effects on the secretion of enzymes, emulsifying agents (e.g., ejection of bile acids from the gall bladder), and motility factors that promote digestion and absorption of nutrients. Insulin and glucagon, secreted from cells in the pancreatic islets, play central roles in the control of blood glucose levels through regulation of glucose entry into cells and the laying down or mobilization of carbohydrate stores in the liver.

▶ **HORMONAL FACTORS IN GUT FUNCTION CONTROL**

In addition to the neural control of the gut, there are diffusible factors that are carried by the bloodstream between parts of the gut and from the gut to the rest of the body. These factors all have a role in regulation of digestion and appetite. Those acting locally, i.e., within the abdomen, are mainly concerned with controlling secretion and motility of the affected gut segment. Those acting further afield are concerned with energy utilization and appetite.

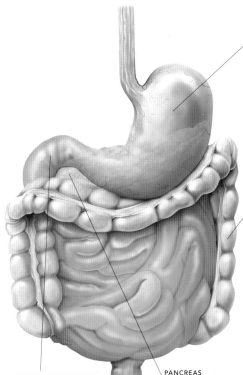

STOMACH

Ghrelin secreted by the stomach acts on the hypothalamus to affect hunger and appetitive behavior

Gastrin secreted in response to the presence of food increases stomach acid secretion

COLON

Glucagon-like peptides 1 and 2 (GLP-1, GLP-2) are active in the colon. GLP-1 suppresses glucagon secretion and restores the sensitivity of pancreatic beta cells. GLP-2 promotes intestinal growth and function

Oxyntomodulin suppresses appetite

Pancreatic peptide YY3-36 suppresses appetite

SMALL INTESTINE (DUODENUM)

Cholecystokinin induces gall bladder contraction to release bile, modulates gastrointestinal motility and increases exocrine pancreas secretion

Secretin regulates water balance and pH of the duodenum by inhibiting gastric acid secretion and increasing bicarbonate secretion by the pancreas

PANCREAS

Insulin and glucagon regulate glucose homeostasis

Pancreatic polypeptide inhibits exocrine pancreas secretion that is caused by cholecystokinin

Structure & function of the small intestine

THE SMALL INTESTINE IS THE PART OF THE GUT TUBE THAT PERFORMS MOST OF THE ABSORPTION OF NUTRIENTS. It must also complete some of the digestive process that has begun in the oral cavity and stomach. The small intestine is divided into three parts: duodenum, jejunum, and ileum. All three parts have a broadly similar structure that optimizes the internal surface area available for the absorption of nutrients.

The inner lining of the gut tube (the mucosa) is thrown into folds called plicae circulares. At a finer scale, the surface of each plica has tiny fingerlike projections called villi. Then, at an ultrastructure level, the individual cells of the mucosa (the enterocytes) have tiny fingerlike extensions called microvilli on their cell surface facing the gut interior to increase cell surface area.

Underlying the mucosa is the muscularis mucosae, a smooth muscle layer that helps to move the mucosal lining into contact with food in the cavity of the gut tube. Like other parts of the gut, the small intestine has layers of smooth muscle in inner circular and outer longitudinal layers. These are responsible for moving the gut contents downstream toward the anus in a process known as peristalsis.

The wall of the small intestine also contains glands for gut secretions and nerve networks to regulate gut function (the enteric nervous system). Finally, the downstream part of the small intestine (the ileum) has abundant immune system cells in its wall (Peyer's patches) that assist in the control of the gut bacteria.

▶ **STRUCTURE OF A TYPICAL LENGTH OF SMALL INTESTINE**
The wall of the small intestine has the same basic structure as other parts of the gut tube, namely a division into an innermost mucosa, submucosa, muscular layer, and an outer serosa. The small intestine has a rich vascular supply, with arteries and veins reaching the gut by passing between the layers of the mesentery that supports the gut.

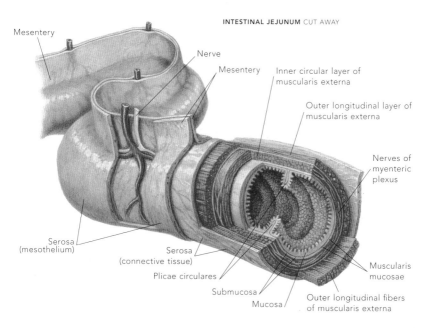

Mesentery

Nerve

Mesentery

Inner circular layer of muscularis externa

Outer longitudinal layer of muscularis externa

Nerves of myenteric plexus

Serosa (mesothelium)

Serosa (connective tissue)

Plicae circulares

Submucosa

Mucosa

Muscularis mucosae

Outer longitudinal fibers of muscularis externa

CELIAC DISEASE

Celiac disease is an autoimmune disorder of the small intestine, meaning that the body's immune system mounts an immune assault on the lining of the gut. It has a genetic basis and may occur in affected people from infancy to adulthood. The immune response is caused by exposure to gluten protein found in wheat, barley, and rye. When gluten protein is acted upon by an enzyme in the gut wall, a cross-reaction by the immune system occurs with tissue of the small intestinal wall, leading to small bowel inflammation and damage to the mucosa. Symptoms and signs include abdominal pain, constipation or diarrhea, failure of children to grow, anemia (reduced circulating hemoglobin), and fatigue. Vitamin deficiencies arise because of the inability of the damaged small intestine to absorb nutrients.

Liver, gall bladder, & bile in digestion

THE LIVER IS A VERY LARGE ORGAN (up to 11 lb/5 kg in large men) that performs myriad metabolic functions. It is primarily a metabolic organ responsible for storing and releasing carbohydrates as required, performing many key aspects of fat metabolism, removing amino groups from amino acids, and converting toxic ammonia to urea for excretion. It also produces the plasma proteins, e.g., albumin, which maintain the osmotic pressure of the blood and transport vital micronutrients like iron and copper around the body. Its other key roles include the production of clotting factors that allow hemostasis in the event of bleeding and the detoxification of foreign chemicals, e.g., alcohol, microbial toxins, and plant alkaloids that are absorbed in the food.

The liver's role in digestion is to produce bile salts for the emulsification of fats. This is a process that breaks up globules of fat that are naturally insoluble in the watery environment of the gut into small particles that can be acted upon by the lipase enzymes of the gut wall. Between meals, bile is stored in the gall bladder, a sac-like organ underneath the liver. When partially digested food (chyme) enters the duodenum from the stomach, cholecystokinin is released from the mucosa of the duodenum to flow through the blood and trigger the ejection of bile from the gall bladder. This bile flows down the bile duct to the duodenum, where it can act on fats.

JAUNDICE

Jaundice is the presence of bile pigment and/or salts in the bloodstream and their accumulation in the skin and sclera to cause yellowing. Jaundice may be caused by excess pigment production, as in hemolytic anemia (prehepatic jaundice), an inability of the liver to process bile pigments (hepatocellular jaundice), or obstruction of the biliary passages to the duodenum (cholestatic or obstructive jaundice).

► MACROSCOPIC AND MICROSCOPIC STRUCTURE OF THE LIVER

The liver is the largest gland in the body and has diverse metabolic functions. Macroscopically, the liver is divided into left and right lobes by the falciform ligament (top). Microscopically (bottom), the liver is composed of many thousands of hexagonal prism-shaped structures called hepatic lobules. The edges of these prisms have a group of structures called the portal triad, consisting of an interlobar bile duct (green), a branch of the portal vein (blue), and a branch of the hepatic artery (red). The center of each lobule has a central vein that drains into the inferior vena cava.

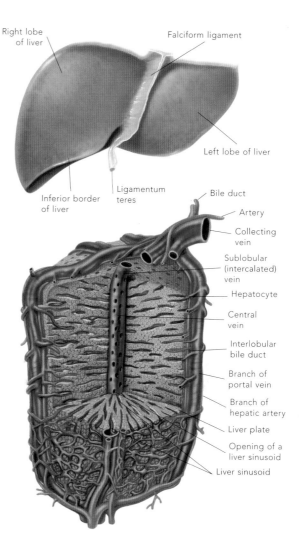

Right lobe of liver

Falciform ligament

Left lobe of liver

Inferior border of liver

Ligamentum teres

Bile duct

Artery

Collecting vein

Sublobular (intercalated) vein

Hepatocyte

Central vein

Interlobular bile duct

Branch of portal vein

Branch of hepatic artery

Liver plate

Opening of a liver sinusoid

Liver sinusoid

Exocrine pancreas in digestion

THE PANCREAS IS AN ABDOMINAL ORGAN THAT CONTAINS BOTH AN ENDOCRINE COMPONENT (the islets of Langerhans) that produces hormones such as insulin and glucagon and an exocrine component (pancreatic acini) that secretes enzymes and bicarbonate ions into the interior of the duodenum. The exocrine pancreas has a duct system (main and accessory pancreatic ducts) that opens through a smooth muscle sphincter (the sphincter of Oddi) into the duodenal cavity at the greater duodenal papilla.

The enzymes that are produced by the pancreas include trypsin, chymotrypsin, and carboxypeptidases, which break protein down into smaller molecules; lipases that digest fats to fatty acids and glycerol; nucleases that digest nucleic acids; and amylases that break down starches to smaller carbohydrates. The bicarbonate ions in pancreatic secretions make the pancreatic juices alkaline (pH of 8.0) and help to neutralize the acid of the stomach so that the pH of the small intestine is returned to neutral. This is important for the proper activation and activity of intestinal and pancreatic enzymes.

Secretion of pancreatic products is under the control of hormones (cholecystokinin and secretin) produced by the small intestine when the stomach juices (chyme) reach the duodenum. Cholecystokinin induces the secretion of enzyme-rich pancreatic juice, whereas secretin causes the release of bicarbonate ion–rich pancreatic juice.

▶ **MACROSCOPIC AND MICROSCOPIC STRUCTURE OF THE PANCREAS**

The pancreas has exocrine (pancreatic acinar cells) and endocrine (alpha, beta, and delta cells of pancreatic islets) components (bottom illustration). Secretions of the exocrine acinar cells reach the duodenum through the main and accessory pancreatic ducts (top illustration).

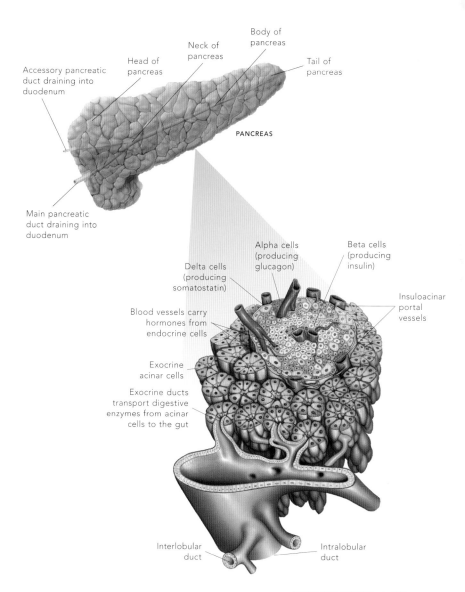

Accessory pancreatic duct draining into duodenum

Head of pancreas

Neck of pancreas

Body of pancreas

Tail of pancreas

PANCREAS

Main pancreatic duct draining into duodenum

Delta cells (producing somatostatin)

Alpha cells (producing glucagon)

Beta cells (producing insulin)

Blood vessels carry hormones from endocrine cells

Insuloacinar portal vessels

Exocrine acinar cells

Exocrine ducts transport digestive enzymes from acinar cells to the gut

Interlobular duct

Intralobular duct

Absorption in the small intestine

THE SMALL INTESTINE IS OPTIMIZED FOR ABSORPTION OF NUTRIENTS BY HAVING A LARGE INTERNAL SURFACE AREA. This is achieved by mucosal folds (plicae circulares) at the macroscopic level and fingerlike projections at the tissue and cellular levels (villi and microvilli, respectively). Another key attribute of the small intestine, particularly in the jejunum, where most absorption occurs, is the presence of a rich vascular supply and a dense capillary bed.

Sugars and amino acids are absorbed through the intestinal wall by active transport, meaning that we actually expend energy to take on nutrients. Small sugar molecules (glucose and fructose) as well as amino acids from the digestion of proteins are carried by the portal venous circulation back to the liver, where the absorbed micronutrients can be processed.

Fats are handled differently because of their lipid solubility and larger size, and they are absorbed passively across the intestinal wall by a process of diffusion. Small fatty acids can be carried to the liver by the portal venous circulation. If the fats are large molecules, they are packaged with special carrier proteins to form chylomicrons, which are carried away from the gut by lymphatic channels (lacteals), which are also prominent in the villi of the intestinal wall. Fat-soluble vitamins (A, K, and D) are also absorbed along with fats.

▶ **ABSORPTION OF NUTRIENTS**

Nutrients in the gut lumen are absorbed across the intestinal mucosa and carried away from the gut by portal venous blood to the liver (sugars, amino acids, nucleic acids, and short-chain fatty acids) or intestinal lymphatics (larger fats).

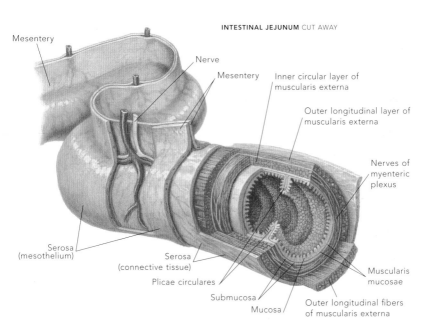

INTESTINAL JEJUNUM CUT AWAY

Mesentery

Nerve

Mesentery

Inner circular layer of muscularis externa

Outer longitudinal layer of muscularis externa

Nerves of myenteric plexus

Serosa (mesothelium)

Serosa (connective tissue)

Plicae circulares

Submucosa

Mucosa

Muscularis mucosae

Outer longitudinal fibers of muscularis externa

Structure & function of the large intestine

THE FUNCTION OF THE LARGE INTESTINE IS TO ABSORB WATER AND MINERALS FROM THE REMNANTS OF THE GUT CONTENTS, as soon as digestion and absorption of useful nutrients are finished. The large intestine mixes waste products, e.g., undigested cellulose and heme pigments, with mucus to assist movement of the stool and excretes the waste as feces through the anus.

The terminal part of the small intestine (the ileum) opens into the initial part of the large intestine (the cecum) through the ileocecal valve. The vermiform appendix is attached to the inner side of the cecum and has a tiny orifice that opens into the cecal cavity. Obstruction of the orifice of the appendix by hard feces or intestinal parasites can damage the appendicular mucosa and lead to appendicitis.

The cecum leads to the ascending colon, which runs upward toward the liver and turns (at the hepatic flexure) to become the transverse colon. The transverse colon runs across the abdomen to the spleen to bend (at the splenic flexure) and run downward as

▶ MACROSCOPIC STRUCTURE OF THE LARGE INTESTINE

The large intestine is arranged in a frame around the small intestine. The initial part of the large intestine is the cecum, which receives the terminal ileum. The cecum has the tiny vermiform appendix attached to it. The ascending colon passes up the right side of the abdomen to bend at the liver (hepatic flexure) to become the transverse colon, which bends at the spleen (splenic flexure) to become the descending colon. This in turn becomes the sigmoid colon, rectum, and anus.

the descending colon. At the pelvic brim, the descending colon becomes the sigmoid colon (so called because of its "S" shape), which becomes the rectum at the middle of the sacrum. The rectum can store feces for short periods until defecation is appropriate. When that is possible, the feces move downward through the anus to the external environment.

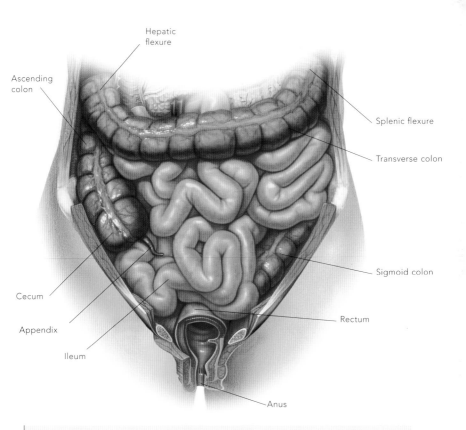

Hepatic flexure

Ascending colon

Splenic flexure

Transverse colon

Sigmoid colon

Cecum

Rectum

Appendix

Ileum

Anus

APPENDICITIS

Appendicitis is an inflammatory condition of the appendix with serious consequences. The process may begin with damage to the mucosa of the appendix that spreads through the wall, causing coagulation of blood in the appendicular vein, loss of effective venous drainage, venous gangrene and possibly rupture. Rupture of the appendix can have serious consequences because bowel contents and gut bacteria may spread throughout the peritoneal cavity leading to bacterial peritonitis. Surgical removal is the standard treatment.

Overview of urinary tract & kidney structure: macroscopic

THE URINARY SYSTEM CONSISTS OF THE TWO KIDNEYS ON THE POSTERIOR ABDOMINAL WALL, paired tubular structures (the ureters) to carry urine to the urinary bladder for storage, and a midline urethra to carry urine from the urinary bladder to the external environment.

The kidney consists of an outer cortex surrounded by a connective tissue capsule and an inner medulla. Ultrafiltration of the blood occurs in the cortex, and some of the tubular network of the nephron (proximal and distal convoluted tubules) is also located here. The renal medulla is a zone where high osmotic pressure is developed by pumping of ions in the long loops of Henle. This osmotic gradient allows the medulla to be used as a zone where water retention from the filtered urine can be adjusted according to need.

Urine is collected in the minor calyces, which combine to form major calyces, which in turn combine to form the renal pelvis. The renal pelvis and calyceal tree are located in the cleft of the renal sinus in the center of the kidney. The renal pelvis continues as the ureter, which runs down the posterior abdominal wall to the bladder.

The urinary bladder is a pelvic organ in both sexes and is essentially a muscular bag that can fill to a volume of 0.6 to 1.1 pints (300 to 500 ml) and empty under voluntary control. The urethra is short in females—only 1.5 inches (4 cm) long, but longer in males—up to 8 inches (20 cm)—and linked with the reproductive system. This means that urinary tract infections of the bladder are much more common in women than men because it is easier for bacteria to ascend the short female urethra.

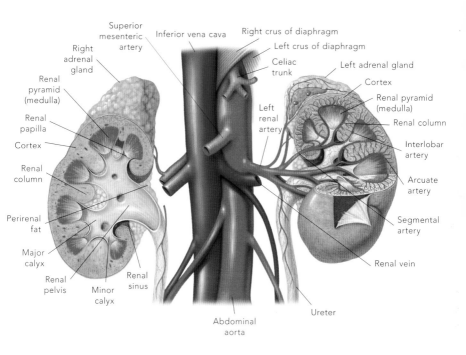

Superior mesenteric artery
Inferior vena cava
Right crus of diaphragm
Left crus of diaphragm
Right adrenal gland
Celiac trunk
Left adrenal gland
Renal pyramid (medulla)
Cortex
Left renal artery
Renal pyramid (medulla)
Renal papilla
Renal column
Cortex
Interlobar artery
Renal column
Arcuate artery
Perirenal fat
Segmental artery
Major calyx
Renal vein
Renal pelvis
Minor calyx
Renal sinus
Ureter
Abdominal aorta

▲ OVERVIEW OF THE KIDNEYS AND THEIR BLOOD VESSELS

This illustration shows both kidneys viewed from the front. The front of the left kidney has been removed to show the vascular branches supplying the cortex. The left kidney has been cut in half to reveal the distinction between the renal cortex, medullary or renal pyramids, and the calyceal system (minor and major calyces) that carries urine to the ureter.

Overview of urinary tract & kidney structure: glomerular & tubular structure

THE KEY MICROSTRUCTURAL ELEMENT OF THE KIDNEY IS THE NEPHRON. This is a functional unit that consists of several components in series, each with a different structure. The first part of the nephron is the glomerulus, where ultrafiltration of the blood occurs. Glomeruli are about 150 to 250 μm (micrometers) in diameter and are found exclusively in the cortex of the kidney.

The next part of the nephron is the proximal convoluted tubule, which is also situated in the renal cortex and is responsible for reabsorbing most of the water in the filtrate and all the useful nutrients (glucose and amino acids). After the proximal convoluted tubule comes the loop of Henle, which is a hairpin-shape looped tube that descends into the renal medulla before coming back up again into the renal cortex. The loop of Henle has thick and thin segments with different functions (see pp. 272–273), but the main function is to regulate water balance in the body.

The next section is the distal convoluted tubule situated within the renal cortex. The early distal tubule is structurally and functionally similar to the loop of Henle, whereas the late distal tubule is similar to the next part of the nephron, the cortical collecting duct. The late distal tubule and cortical collecting duct are under hormonal influence (aldosterone, antidiuretic hormone, and atrial natriuretic peptide) to fine-tune water, acid-base, and electrolyte balance.

Finally, the urine passes down the medullary collecting system to the apex of the medullary pyramids. Water may still be absorbed from the medullary collecting system under the influence of antidiuretic hormone.

▶ **MICROSTRUCTURE OF THE NEPHRON**

The nephron consists of a glomerulus, proximal convoluted tubule, loop of Henle, distal convoluted tubule, and collecting ducts. The entire system of the glomerulus and tubules is richly vascularized.

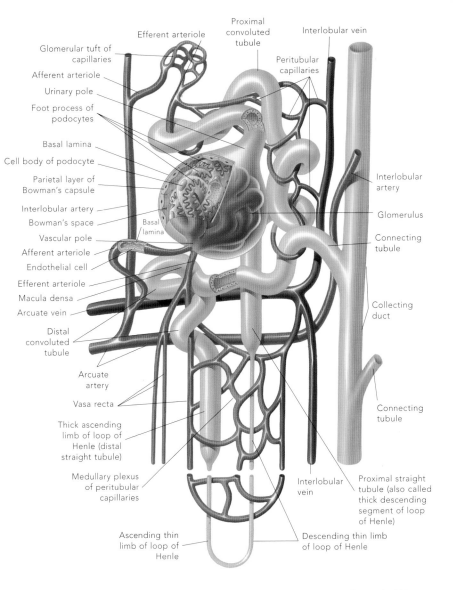

Efferent arteriole

Glomerular tuft of capillaries

Afferent arteriole

Urinary pole

Foot process of podocytes

Basal lamina

Cell body of podocyte

Parietal layer of Bowman's capsule

Interlobular artery

Bowman's space

Vascular pole

Afferent arteriole

Endothelial cell

Efferent arteriole

Macula densa

Arcuate vein

Distal convoluted tubule

Arcuate artery

Vasa recta

Thick ascending limb of loop of Henle (distal straight tubule)

Medullary plexus of peritubular capillaries

Ascending thin limb of loop of Henle

Proximal convoluted tubule

Interlobular vein

Peritubular capillaries

Basal lamina

Interlobular artery

Glomerulus

Connecting tubule

Collecting duct

Connecting tubule

Interlobular vein

Proximal straight tubule (also called thick descending segment of loop of Henle)

Descending thin limb of loop of Henle

Glomerulus & ultrafiltration

THE GLOMERULUS CONSISTS OF
AN INNER TUFT OF GLOMERULAR
CAPILLARIES COVERED BY PODOCYTES
(foot process cells) of the inner
or visceral layer of the glomerular
capsule. This is surrounded by a space
(glomerular, or Bowman's capsule)
from which the proximal convoluted
tubule arises. The outer part of the
glomerulus is the outer or parietal layer
of the glomerular capsule.

Glomerular filtration is a passive
and nonselective process that separates
the fluid component of blood (with
dissolved sugars, amino acids, and
ions) from the plasma proteins.
Filtration is achieved by the passage
of blood through the glomerular
capillaries. There are three layers of the
filtration membrane. The first is the
fenestrations (tiny windows) in the
glomerular capillary endothelial
cells. The second is a thin layer of
extracellular material (basal lamina)
on which the endothelial cells sit.
The final component is the podocyte
(foot process cell) layer. This is actually
the finest filter because the spaces
between the podocyte processes are

only 6 to 7 nm (i.e., 6 to 7 millionths
of a millimeter) wide.

Clogging of the filtration spaces is
prevented by the presence of a negative
charge on the podocyte processes that
repels proteins and larger molecules.
These three filtration layers effectively
keep cells and plasma proteins like
albumin within the blood of the
glomerular capillaries but permit
125 ml/min of water and dissolved
substances in the ultrafiltrate to flow
into the proximal convoluted tubules.

▶ **THE PROCESS OF
ULTRAFILTRATION**

Ultrafiltration depends on a pressure
gradient between the afferent glomerular
arteriole (glomerular blood hydrostatic
pressure or GBHP of 55 mm Hg) tending
to force fluid out through the filtration
spaces and the combined capsular
hydrostatic and blood osmotic pressures
(15 and 30 mm Hg, respectively) tending
to oppose filtration. The net filtration
pressure producing a filtrate that leaves
the glomerulus through the urinary pole
is therefore 10 mm Hg.

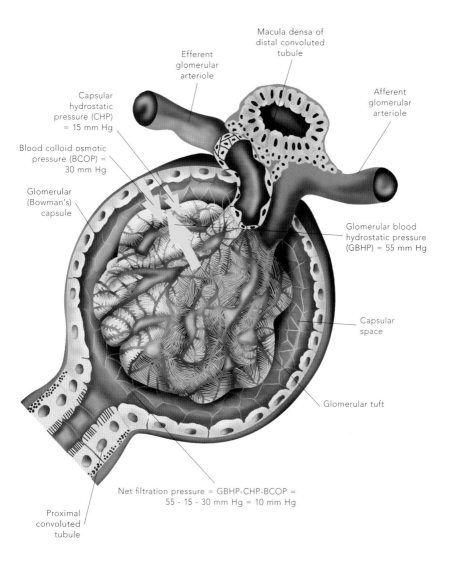

Macula densa of distal convoluted tubule

Efferent glomerular arteriole

Afferent glomerular arteriole

Capsular hydrostatic pressure (CHP) = 15 mm Hg

Blood colloid osmotic pressure (BCOP) = 30 mm Hg

Glomerular (Bowman's) capsule

Glomerular blood hydrostatic pressure (GBHP) = 55 mm Hg

Capsular space

Glomerular tuft

Net filtration pressure = GBHP-CHP-BCOP = 55 - 15 - 30 mm Hg = 10 mm Hg

Proximal convoluted tubule

Tubular reabsorption & secretion

REABSORPTION OF ALMOST ALL OF THE WATER—and all of the essential nutrients—from the glomerular filtrate occurs in the tubular components of the nephron. There is also some tubular secretion of hydrogen and potassium ions, creatinine, and some drugs. Tubular absorption and secretion depends on the transfer of these substances between the tubules and the rich vascular supply of the renal cortex and medulla.

Approximately 65% of the water, sodium, potassium, calcium, chloride, and magnesium ions in the filtrate are absorbed in the proximal convoluted tubule, along with nearly all the glucose and amino acids. About 90% of bicarbonate ions are also absorbed at this site. Some secretion also occurs at the proximal convoluted tubules (hydrogen ions, nitrogenous waste like uric acid, and some drugs). The loop of Henle is a site for reabsorption of 20% of water from the filtrate (at the thin-walled descending part) and 25% of the sodium and chloride (at the thick ascending part). The distal convoluted tubule and collecting duct

are responsible for reabsorption of most of the remaining water and nearly all of the remaining sodium, chloride, bicarbonate, and calcium ions. The distal convoluted tubule and collecting duct is also a site for potassium and hydrogen ion secretion, under hormonal regulation. Some drugs are also secreted here.

The end result is that 124 ml of the 125 ml/min glomerular filtrate, plus the bulk of useful nutrients and minerals, are reabsorbed by the time that the apex of the renal pyramid is reached. Nitrogenous waste like urea, creatinine, ammonia, and uric acid are therefore the major constituents of urine.

▶ **STRUCTURE AND FUNCTION OF THE TUBULAR SYSTEM**

The tubular components of the nephron include the proximal convoluted tubule for absorption of most nutrients and ions and water, the loop of Henle for some water and ion reabsorption, the early distal convoluted tubule for hormone-regulated absorption and secretion of remaining ions, and the late distal convoluted tubule and collecting ducts for water reabsorption.

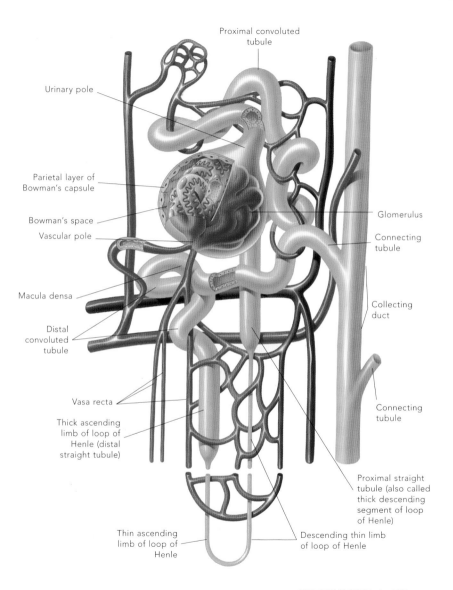

Proximal convoluted tubule

Urinary pole

Parietal layer of Bowman's capsule

Bowman's space

Vascular pole

Macula densa

Distal convoluted tubule

Vasa recta

Thick ascending limb of loop of Henle (distal straight tubule)

Thin ascending limb of loop of Henle

Glomerulus

Connecting tubule

Collecting duct

Connecting tubule

Proximal straight tubule (also called thick descending segment of loop of Henle)

Descending thin limb of loop of Henle

Measuring urine chemistry & renal function

NORMAL URINE IS THE FLUID THAT REMAINS ONCE ALL THE FILTRATION, absorption, and secretion of the kidney is completed. About 125 ml/min is filtered through the glomeruli (glomerular filtration rate, or GFR) and 124 ml/min of that is reabsorbed in the tubules and collecting ducts, so that about 1 ml/min passes into the bladder.

Most people produce between 2.1 and 3.8 pints (1.0 and 1.8 L) of urine each day. Urine is a yellow fluid due to the pigment urochrome, which is a breakdown product of hemoglobin. Urine should be clear because it should never contain more than trace amounts of protein and it should be diluted enough to read through. Normal urine is also sterile, i.e., no bacteria should be present. Cloudy urine may mean that there is infection in the urinary tract or that plasma proteins are leaking through the filtration membrane in the glomerulus (nephrotic syndrome). The pH of the urine should be slightly acidic (about 6.0, but within the range of 4.5 to 8.0).

Poor renal function can occur when the amount of blood reaching the kidneys is reduced or when kidney tissue is lost. Signs of poor renal function include reduced or increased urine output (depending on the type of renal problem), puffy eyes and swollen hands and feet, increased blood pressure, and increased concentration of urea and creatinine in the blood. Unfortunately, there are often few obvious symptoms until advanced renal failure has been reached.

▶ **NORMAL VALUES OF URINE AND BLOOD CHEMISTRY**
Urine and blood chemistry provide a good indication of renal function. Poor renal function is indicated by elevations in urea and creatinine in the blood. Changes in bicarbonate ion concentration and pH indicate how well the kidney is regulating acid-base balance. Urine should have minimal protein content and very few cells.

NORMAL VALUES OF URINE CHEMISTRY

Osmotic concentration:	850 to 1,340 mOsm/L
Specific gravity:	1.003 to 1.030
pH:	4.5 to 8.0, with a mean of 6.0
Bacterial content:	Nil, urine should be sterile
Red blood cells:	100/ml
White blood cells:	500/ml
Sodium:	330 mg/dl
Potassium:	166 mg/dl
Chloride:	530 mg/dl
Calcium:	17 mg/dl
Urea:	1.8 g/dl
Creatinine:	150 mg/dl
Ammonia:	60 mg/dl
Uric acid:	40 mg/dl
Urobilin (yellow pigment):	125 μg/dl

NORMAL VALUES OF BLOOD PLASMA CHEMISTRY THAT INDICATE RENAL FUNCTION

Sodium:	138 mM
Potassium:	4.4 mM
Chloride:	106 mM
Bicarbonate:	27 mM
pH:	7.35 to 7.45
Urea:	10 to 20 mg/dl
Creatinine:	1 to 1.5 mg/dl
Ammonia:	< 0.1 mg/dl

IMPORTANT NOTE: THESE VALUES ARE PROVIDED FOR EDUCATIONAL PURPOSES ONLY AND SHOULD NOT BE USED FOR SELF-DIAGNOSIS. REFERENCE NORMAL VALUES MAY VARY BETWEEN TESTING LABORATORIES DEPENDING ON THE SAMPLE REFERENCE POPULATION.

Regulation of glomerular filtration rate

THE FILTRATION OF BLOOD THROUGH THE GLOMERULI IS UNDER TIGHT REGULATION, such that a relatively constant glomerular filtration rate is maintained over a wide range of arterial blood pressure. One key element in the autoregulation of the glomerular filtration rate is a zone called the juxtaglomerular apparatus. This includes a structure called the macula densa that is situated at the junction of the ascending limb of the loop of Henle and the distal convoluted tubule. The macula densa is in contact with modified smooth muscle cells (juxtaglomerular cells) of the afferent and efferent glomerular arterioles (the vessels into and out of the glomerulus).

The macula densa senses the flow of glomerular filtrate down the distal tubule by detecting the amount of sodium and chloride ions passing it in a given time. When flow is too high, the macula densa then secretes an unidentified chemical messenger that diffuses locally (a paracrine effect) to cause constriction of the afferent arteriole that carries blood to the

▶ **AUTOREGULATION MECHANISMS**

The glomerular filtration rate (GFR) is regulated to a narrow range (around 125 ml/min) by a combination of a myogenic mechanism at the level of the afferent arteriole and a paracrine (local diffusible messenger) mechanism involving the macula densa.

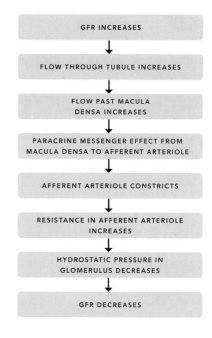

GFR INCREASES

↓

FLOW THROUGH TUBULE INCREASES

↓

FLOW PAST MACULA DENSA INCREASES

↓

PARACRINE MESSENGER EFFECT FROM MACULA DENSA TO AFFERENT ARTERIOLE

↓

AFFERENT ARTERIOLE CONSTRICTS

↓

RESISTANCE IN AFFERENT ARTERIOLE INCREASES

↓

HYDROSTATIC PRESSURE IN GLOMERULUS DECREASES

↓

GFR DECREASES

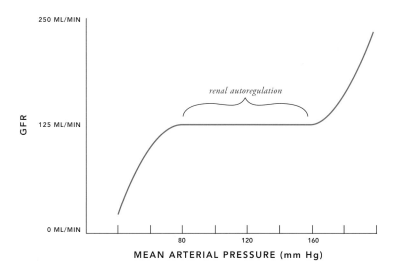

GFR

250 ML/MIN

renal autoregulation

125 ML/MIN

0 ML/MIN

80   120   160

MEAN ARTERIAL PRESSURE (mm Hg)

glomerulus. When the smooth muscle of the afferent arteriole constricts, the vascular resistance rises in the arteriole and the flow of blood to the glomerulus drops, thereby also lowering the volume of filtrate produced. This reflex is known as tubuloglomerular feedback.

Autoregulation of glomerular filtrate may also occur by a myogenic mechanism: when arterial pressure increases, the wall of the afferent glomerular arteriole is stretched and the smooth muscle in the wall contracts, constricting the afferent arteriole and reducing filtrate production.

Regulation of potassium

POTASSIUM (K⁺) IS AN IMPORTANT ION WITHIN THE CELLS OF THE BODY. It is also present in the extracellular fluid in small amounts, but its concentration there must not rise too high or the activity of excitable cells like nerve and cardiac muscle cells will be severely disrupted and death from cardiac arrest may result.

Potassium in the body is in part regulated by a balance between its loss in the glomerular filtrate and its secretion or reabsorption in the tubules of the nephron. About 23 g of potassium is passed into the glomerular spaces each day as part of the glomerular filtrate. Between 60% and 80% of that will be reabsorbed at the proximal convoluted tubules by either crossing the tight junctions between the tubular epithelial cells or passing through the cytoplasm of the tubular cells themselves (mechanism unknown). Another 10% to 30% is reabsorbed at the level of the loop of Henle, although some minor secretion may also occur here.

Potassium may also be secreted at the level of the distal convoluted tubules and cortical collecting ducts under the influence of the steroid hormone aldosterone. The net loss of potassium from the kidneys each day is about 4 g, and this (along with loss in sweat and other glandular secretions) must be made up by ingestion of potassium in food.

▶ **MECHANISMS OF POTASSIUM REGULATION**

About 600 mmol of potassium passes into the glomerular filtrate each day, and most of that is reabsorbed at the level of the proximal convoluted tubules and loop of Henle. Some will be secreted in the distal convoluted tubule under the regulation of the steroid aldosterone. Potassium may be absorbed through the space between proximal convoluted tubule epithelia cells or through the cells themselves by an unknown mechanism. The normal concentration of potassium in the blood is 4.4 mM (range of 3.5 to 5.0 mM).

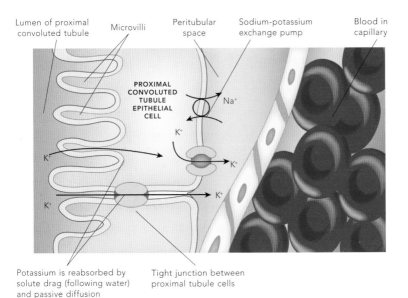

Lumen of proximal convoluted tubule

Microvilli

Peritubular space

Sodium-potassium exchange pump

Blood in capillary

PROXIMAL CONVOLUTED TUBULE EPITHELIAL CELL

Na^+

K^+

K^+

K^+

K^+

K^+

K^+

Potassium is reabsorbed by solute drag (following water) and passive diffusion

Tight junction between proximal tubule cells

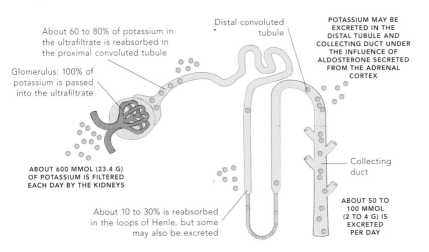

About 60 to 80% of potassium in the ultrafiltrate is reabsorbed in the proximal convoluted tubule

Distal convoluted tubule

POTASSIUM MAY BE EXCRETED IN THE DISTAL TUBULE AND COLLECTING DUCT UNDER THE INFLUENCE OF ALDOSTERONE SECRETED FROM THE ADRENAL CORTEX

Glomerulus: 100% of potassium is passed into the ultrafiltrate

ABOUT 600 MMOL (23.4 G) OF POTASSIUM IS FILTERED EACH DAY BY THE KIDNEYS

Collecting duct

ABOUT 50 TO 100 MMOL (2 TO 4 G) IS EXCRETED PER DAY

About 10 to 30% is reabsorbed in the loops of Henle, but some may also be excreted

Renal regulation of acid-base balance

THE KIDNEYS PLAY A KEY ROLE IN REGULATING THE PH OF THE BLOOD, acting in concert with the regulation of carbon dioxide (CO_2) by lung ventilation. By a combination of interlooped negative feedback pathways, deviations from optimum pH cause responses from both renal and respiratory mechanisms to return to homeostasis. We have already discussed how ventilation can affect pH, but metabolic corrections are also important.

When carbon dioxide combines with water in the blood (a reaction catalyzed in red blood cells by the enzyme carbonic anhydrase—CA), the bicarbonate ion is produced. Bicarbonate ions (HCO_3^-) can be reabsorbed from the proximal convoluted tubule of the kidney in a mechanism that requires hydrogen ions to be first ejected from the tubular cells into the filtrate. These combine with the bicarbonate ion to form carbonic acid (H_2CO_3), which is converted to water and CO_2 by carbonic anhydrase enzyme on the apical membrane of the renal tubule cell.

CO_2 diffuses into the tubule cell, where it is again converted to bicarbonate ions, which are then transported to the interstitial fluid outside the tubule. This mechanism allows the cells of the proximal tubule to absorb approximately 90% of the bicarbonate in the filtrate.

▶ **THE PROCESS OF RENAL REGULATION**

The kidneys control acid-base balance of the body by a process of regulated bicarbonate ion reabsorption in the proximal convoluted tubule. This requires the ejection of hydrogen ions into the tubule to capture carbon dioxide, which is then converted to bicarbonate ions, which in turn are shunted to the interstitial fluid outside the cell and then eventually absorbed into the bloodstream. The process uses large amounts of adenosine triphosphate (ATP).

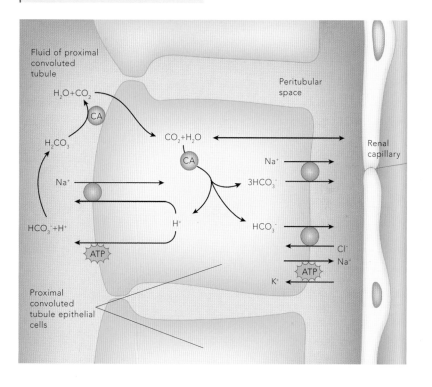

Urine concentration, water conservation, & antidiuretic hormone

THE KIDNEYS ARE VERY EFFECTIVE AT CONSERVING WATER. They can produce urine with a concentration of 1200 mOsm, which is four times the ionic concentration of blood and other tissues. The process of concentrating urine requires the absorption of water by osmosis in the late distal convoluted tubule and collecting duct of the kidney, under the influence of antidiuretic hormone (ADH), otherwise known as vasopressin.

The hormone ADH is produced by the posterior pituitary in response to dehydration, but ADH would be unable to exert its effects if there were not already a strong concentration gradient present in the renal medulla. This gradient is produced by a particular type of nephron called a juxtamedullary nephron. These nephrons have very long loops of Henle that run deep into the renal medulla.

The thick ascending part of the loop of Henle contains very metabolically active cells that pump sodium and chloride ions from the filtrate into the surrounding tissue of the renal medulla. The process of pumping those ions begins to attract water from the filtrate in the thin descending tubule, making it possible to pump even more sodium and chloride out of the thick ascending tubule and achieve even higher concentrations in the surrounding tissues of the medulla.

This boot-strapping process is called a countercurrent multiplier because the descending and ascending tubules have fluid flowing in opposite directions.

▶ **THE ACTION OF ADH ON THE DISTAL TUBULE AND COLLECTING DUCT**

The hormone ADH (also called vasopressin) acts on the late distal tubule/collecting ducts through a G protein–coupled receptor. This promotes incorporation of aquaporin-2 channels into the membrane of the collecting duct cells to allow water to pass into the cells from the filtrate and on via aquaporin-3 channels in the membrane of the other side of the cell into the bloodstream.

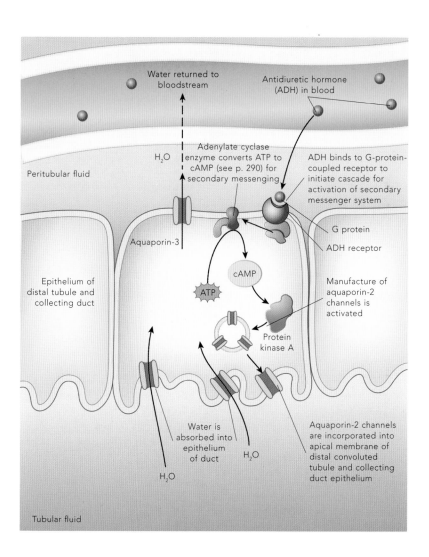

Water returned to bloodstream

Antidiuretic hormone (ADH) in blood

Peritubular fluid

H_2O

Adenylate cyclase enzyme converts ATP to cAMP (see p. 290) for secondary messenging

ADH binds to G-protein-coupled receptor to initiate cascade for activation of secondary messenger system

Aquaporin-3

G protein

ADH receptor

cAMP

ATP

Epithelium of distal tubule and collecting duct

Manufacture of aquaporin-2 channels is activated

Protein kinase A

Water is absorbed into epithelium of duct

H_2O

Aquaporin-2 channels are incorporated into apical membrane of distal convoluted tubule and collecting duct epithelium

H_2O

Tubular fluid

Renin-angiotensin system

THE RENIN-ANGIOTENSIN SYSTEM HAS BEEN COVERED WITH RESPECT TO THE REGULATION OF ARTERIAL BLOOD PRESSURE (see pp. 208–209), but the system is also important in regulation of glomerular filtration.

Low blood pressure is detected by juxtaglomerular cells of the juxtaglomerular apparatus. These cells produce the hormone renin that catalyzes the conversion of angiotensinogen to angiotensin I. Angiotensin I is converted to angiotensin II by angiotensin-converting enzyme in the endothelial cells of the pulmonary vessels. Angiotensin II acts on the adrenal cortex to stimulate release of the steroid hormone aldosterone, which acts on the kidney tubules to increase sodium absorption and the water that follows the sodium by osmosis. This increases blood volume and the systemic arterial pressure, which in turn increases the production of glomerular filtrate.

Angiotensin II also acts on systemic arterioles to make the smooth muscle in their walls contract, increase peripheral resistance, and raise arterial pressure, which also increases glomerular filtration rate. Angiotensin II also promotes vasoconstriction of the arterioles that leave the glomerulus, raising the hydrostatic pressure in the glomerular capillaries to increase glomerular filtration.

▶ **THE RENIN-ANGIOTENSIN SYSTEM**

This flow chart shows the role of the renin-angiotensin system in controlling blood volume and pressure. The system depends on a cascade of enzymes converting substrates. Renin catalyzes the conversion of angiotensinogen to angiotensin I, and angiotensin-converting enzyme catalyzes the conversion of angiotensin I to angiotensin II. The result is retention of sodium (Na^+) followed by osmotic retention of water, as well as an effect on the smooth muscle of arterioles to increase peripheral resistance and raise blood pressure.

RENIN-ANGIOTENSIN SYSTEM

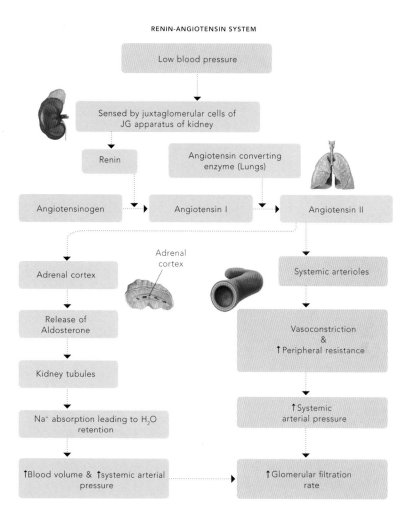

Low blood pressure

Sensed by juxtaglomerular cells of
JG apparatus of kidney

Renin

Angiotensin converting
enzyme (Lungs)

Angiotensinogen

Angiotensin I

Angiotensin II

Adrenal cortex

Adrenal
cortex

Systemic arterioles

Release of
Aldosterone

Vasoconstriction
&
↑Peripheral resistance

Kidney tubules

↑Systemic
arterial pressure

Na+ absorption leading to H₂O
retention

↑Blood volume & ↑systemic arterial
pressure

↑Glomerular filtration
rate

Physiology of micturition

THE PROCESS OF PASSING URINE IS KNOWN AS MICTURITION. When the bladder is full, stretch receptors in the bladder wall signal through the pelvic nerves to the spinal cord parasympathetic nucleus. Reflex activation of the parasympathetic neurons that control the detrusor smooth muscle of the bladder causes transient rises in bladder pressure that force small amounts of urine past the smooth muscle internal bladder sphincter into the upper urethra. This discomfort is interpreted as the signal that the bladder is full.

A sphincter of voluntary skeletal muscle (the external urethral sphincter) surrounds the urethra below the internal sphincter. One can control the further movement of urine for a short time by contracting the external voluntary sphincter, but the need to pass urine will become increasingly urgent as bladder contractions continue. When it is socially appropriate to pass urine, one relaxes the voluntary sphincter, and the pelvic floor muscles that support the bladder neck and the reflexive activation of the detrusor will force urine down the urethra and out to the external environment.

Loss of pelvic floor support and a weakened external urethral sphincter make it difficult to control the flow of urine when the urge arrives. This can be a particular problem for women whose pelvic floor has been damaged from multiple childbirths and who have reached menopause because hormonal support from circulating estrogen is essential for maintaining pelvic floor muscle tone.

▶ **CONTROL OF MICTURITION**
The expulsion of urine from the body requires muscle in the urinary bladder wall (the detrusor) to contract under the influence of the sacral spinal cord reflex centers. The neck of the urinary bladder is surrounded by a smooth muscle sphincter (the internal urethral sphincter) and a skeletal muscle external urethral sphincter around the uppermost urethra (not shown).

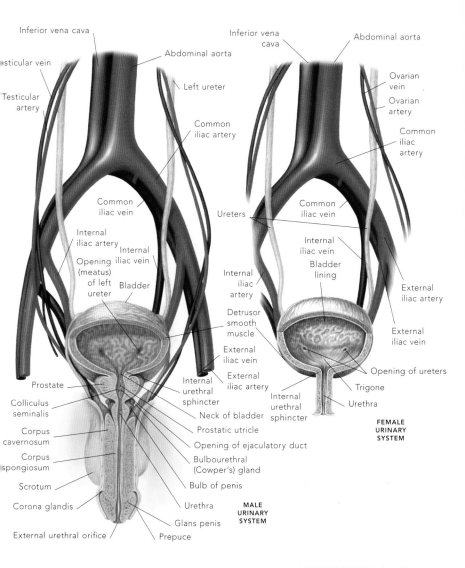

Inferior vena cava

Testicular vein

Testicular artery

Abdominal aorta

Left ureter

Common iliac artery

Common iliac vein

Internal iliac artery

Internal iliac vein

Opening (meatus) of left ureter

Bladder

Prostate

Colliculus seminalis

Corpus cavernosum

Corpus spongiosum

Scrotum

Corona glandis

External urethral orifice

Internal urethral sphincter

Neck of bladder

Prostatic utricle

Opening of ejaculatory duct

Bulbourethral (Cowper's) gland

Bulb of penis

Urethra

Glans penis

Prepuce

Internal iliac artery

Detrusor smooth muscle

External iliac vein

External iliac artery

Ureters

MALE URINARY SYSTEM

Inferior vena cava

Abdominal aorta

Ovarian vein

Ovarian artery

Common iliac artery

Common iliac vein

Internal iliac vein

Bladder lining

External iliac artery

External iliac vein

Opening of ureters

Trigone

Urethra

Internal urethral sphincter

FEMALE URINARY SYSTEM

Overview of the endocrine system

THE ENDOCRINE SYSTEM IS A DIVERSE COLLECTION OF DUCTLESS GLANDS DISTRIBUTED THROUGHOUT THE BODY. The glands communicate with each other and the tissues of the body through diffusible substances called hormones. Hormones may be chains of amino acids (polypeptides); derivatives of amino acids, e.g., thyroid hormone; or steroid molecules, e.g., estrogen, progesterone, and aldosterone. The prostaglandins, which are highly active lipid molecules released from the membranes of many different types of cells, could also be included, but they are usually locally acting molecules that are not secreted from discrete glands.

Hormones are transported by the bloodstream or diffuse through body cavities and tissue spaces to reach a target organ, where they exert an effect. This might be a change in the permeability or electrical state of the plasma membrane, stimulation of the synthesis of proteins or regulatory molecules in the cell, activation or inactivation of enzymes, stimulation of cell division, or the promotion of secretory activity.

Most of the effects of the endocrine system are on reproduction, growth, and development; dealing with stress; maintaining water and electrolyte balance; and regulating nutrient balance in the blood.

A key feature of the function of hormones and the endocrine system is the negative feedback cycle, in which a hormone and/or its physiological effects inhibit the chemical signals that drive hormone production and thus return the internal body state to normal (homeostasis).

▶ **ENDOCRINE GLANDS OF THE BODY**

The endocrine glands are located close to the midline of the body. The master gland is the pituitary, situated immediately below the brain. The thyroid and parathyroids are located in the neck. The thymus (actually a lymphoid and endocrine gland) is in the chest cavity, and the adrenals and endocrine pancreas are in the abdominal cavity. Finally, the gonads (testes in males and ovaries in females) are endocrine glands.

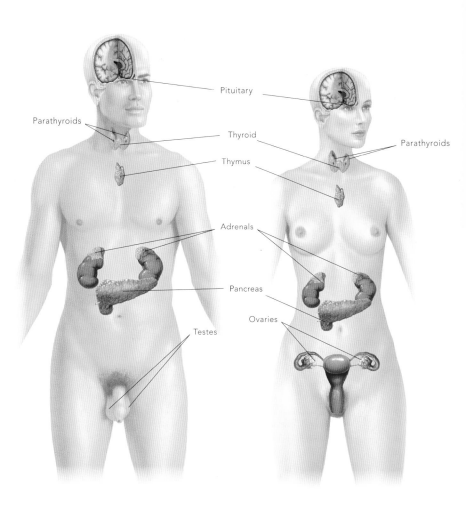

Pituitary

Parathyroids

Thyroid

Parathyroids

Thymus

Adrenals

Pancreas

Ovaries

Testes

Cellular action of hormones

NONSTEROIDAL AND WATER-SOLUBLE PEPTIDE HORMONES ARE UNABLE TO ENTER CELLS. This is because they cannot cross the bilipid layers of the plasma (cell) membrane. Instead, they must act on the cell indirectly by binding to receptors on the cell surface. This is known as a second-messenger system because the binding of the hormone to the receptor in the cell membrane sets up a cascade of events that involves additional messenger molecules.

When binding occurs, the activated receptor sets off a biochemical cascade that activates an enzyme. The enzyme catalyzes reactions that produce secondary messenger molecules such as cyclic AMP (cAMP or cyclic adenine monophosphate) from adenosine triphosphate (ATP) in the cytoplasm. There are other possible secondary messengers, e.g., cyclic guanosine triphosphate (cGTP) associated with G protein–coupled receptors and even simple calcium ions. The secondary messenger then goes on to act on a variety of cellular processes.

Steroid hormones behave differently because they are lipid-soluble and can diffuse through the plasma membrane of cells. Once inside the cell, steroids can bind to a receptor in the cytoplasm that allows them to cross the nuclear membrane. When inside the nucleus, the steroid can bind to a specific nuclear receptor protein. The hormone nuclear receptor complex then binds to specific sites on the cell's DNA to activate genes and induce a change in cell metabolism through the production of proteins.

▶ **MECHANISMS OF ACTION OF HORMONES**

The mechanisms of action of steroidal and nonsteroidal hormones are quite distinct. Steroidal hormones are able to enter the cell and bind to intracytoplasmic receptors, which translocate to the nucleus to affect protein synthesis. Nonsteroidal (peptide or amino acid) hormones act via receptors at the cell surface. Binding of the peptide hormone to the receptor activates an enzyme cascade that generates secondary messengers like cAMP to affect cellular function.

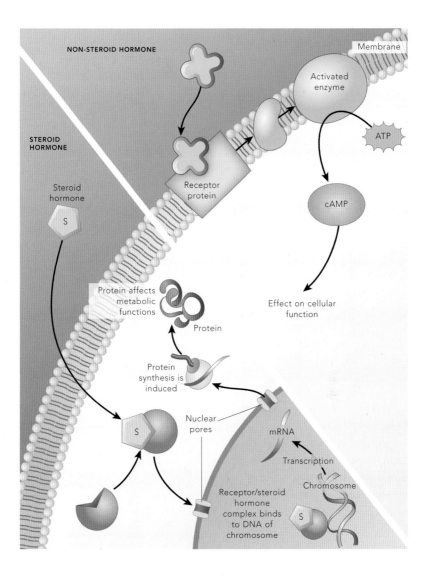

NON-STEROID HORMONE

Membrane

Activated enzyme

ATP

STEROID HORMONE

Steroid hormone

S

Receptor protein

cAMP

Protein affects metabolic functions

Protein

Effect on cellular function

Protein synthesis is induced

Nuclear pores

mRNA

S

Transcription

Chromosome

Receptor/steroid hormone complex binds to DNA of chromosome

S

The anterior pituitary gland & its hormones

THE PITUITARY GLAND IS A PEA-SIZED GLAND SITUATED IMMEDIATELY BENEATH THE HYPOTHALAMUS OF THE BRAIN. It is connected to the hypothalamus by a stalk and is divided into anterior and posterior parts. The anterior pituitary (adenohypophysis) receives venous blood from capillary networks that start in the hypothalamus and collect together into tiny portal veins that run down the pituitary stalk. This hypophyseal or pituitary portal system carries releasing or inhibitory factor hormones produced by groups of nerve cells in the hypothalamus down to the endocrine cells in the anterior pituitary.

All pituitary hormones are peptides that act through secondary messenger systems and are mostly regulated by negative feedback. The hormones released from the anterior pituitary include thyroid-stimulating hormone that acts on the thyroid gland to stimulate thyroxine and triiodothyronine production, adrenocorticotropic hormone (ACTH) that acts on the adrenal cortex to stimulate production of corticosteroids and sex steroids, and growth hormone that acts on diverse parts of the body (bones, muscles, and internal organs) to promote growth.

Other hormones include follicle-stimulating hormone and luteinizing hormone that act on the ovaries or testes to regulate sex cell generation and steroid hormone production, melanocyte-stimulating hormone to increase melanin production in the skin, and prolactin to promote the production of milk in the mammary glands. These hormones are carried from the anterior pituitary by the veins draining the base of the brain and circulate throughout the body.

▶ **MECHANISM OF ACTION OF THE ANTERIOR PITUITARY**

The cells of the anterior pituitary produce a variety of hormones with diverse effects throughout the body. The anterior pituitary cells that produce these hormones are themselves regulated by releasing and inhibitory factors secreted by the cells of the pituitary. These factors pass down to the anterior pituitary by the blood flowing down the pituitary portal system.

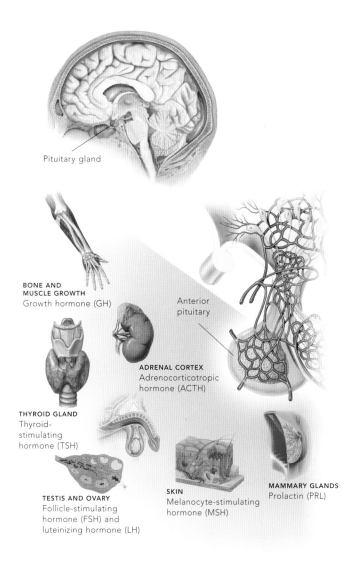

Pituitary gland

BONE AND MUSCLE GROWTH
Growth hormone (GH)

Anterior pituitary

ADRENAL CORTEX
Adrenocorticotropic hormone (ACTH)

THYROID GLAND
Thyroid-stimulating hormone (TSH)

TESTIS AND OVARY
Follicle-stimulating hormone (FSH) and luteinizing hormone (LH)

SKIN
Melanocyte-stimulating hormone (MSH)

MAMMARY GLANDS
Prolactin (PRL)

The posterior pituitary gland & its hormones

THE POSTERIOR PITUITARY GLAND (neurohypophysis) is under neural control from the brain by axonal pathways (hypothalamo-neurohypophyseal tract) starting from neurosecretory cells in the supraoptic and paraventricular nuclei of the hypothalamus. The axons of these cells terminate in the posterior pituitary on capillary beds so that hormones can be directly released into the bloodstream there.

In effect, the posterior pituitary is not really an endocrine gland because it doesn't actually make hormones—it is simply a storage area for hypothalamic hormones. Oxytocin is released during childbirth to induce contraction of the uterus to expel the fetus and placenta through the birth canal. Synthetic oxytocin is routinely used to stimulate labor when delivery of the fetus is necessary. Oxytocin is also important for the milk ejection reflex. When an infant suckles the nipple, oxytocin is released to induce contraction of myoepithelial cells around the mammary gland lobules to squeeze milk out.

DIABETES INSIPIDUS

Diabetes insipidus is a condition in which there is excessive thirst and the secretion of large amounts of dilute urine. Reduction of fluid intake has no effect on urine output. The condition may be neurogenic or central, i.e., due to inadequate secretion of ADH (vasopressin) from the posterior pituitary. It may also be nephrogenic, due to the inability of the kidney to respond to ADH. Patients with diabetes insipidus may survive for many years by consuming large amounts of water, but additional problems like the loss of potassium may eventually have serious consequences. Neurogenic diabetes insipidus can be treated by supplementation with desmopressin. Nephrogenic diabetes insipidus can be treated with some diuretics that shift water reabsorption toward the proximal convoluted tubule of the nephron.

▶ MECHANISM OF ACTION OF THE POSTERIOR PITUITARY

Hormones of the posterior pituitary are made by the neurosecretory neurons of the hypothalamus and travel down the hypothalamo-neurohypophyseal tract to be released from axon terminals in the posterior pituitary. Oxytocin and ADH (vasopressin) are the two hormones found in the posterior pituitary.

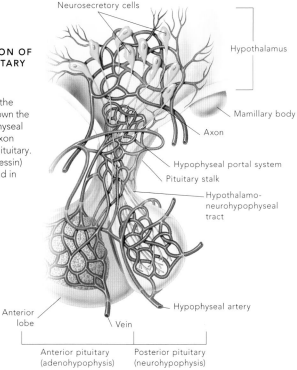

Neurosecretory cells

Hypothalamus

Mamillary body

Axon

Hypophyseal portal system

Pituitary stalk

Hypothalamo-neurohypophyseal tract

Hypophyseal artery

Anterior lobe

Vein

Anterior pituitary (adenohypophysis)

Posterior pituitary (neurohypophysis)

Antidiuretic hormone (ADH), also known as vasopressin, inhibits or prevents urine production by stimulating the reabsorption of water from the filtrate passing down the collecting ducts of the renal medulla. ADH therefore increases the blood volume and also can act on vascular smooth muscle to cause constriction of arterioles, hence its other name of vasopressin. Drinking alcohol inhibits ADH production and leads to a diuresis. Impaired secretion of ADH causes a condition known as diabetes insipidus, in which large amounts of water are lost from the body.

Pineal gland & melatonin

THE PINEAL GLAND IS PART OF THE EPITHALAMUS OF THE BRAIN, and the gland and its hormone melatonin play an important role in the timing of seasonal reproduction in many mammals. Information about the length of the day appears to be transmitted to the rest of the brain by the pattern of secretion of melatonin. The pineal gland secretes melatonin —known as a neurohormone—in response to light and dark cycles and regulates the sleep/wake cycle. Information about light levels comes from the retina through retinohypothalamic pathways to the suprachiasmatic nucleus of the hypothalamus, which in turn influences pineal gland activity.

In humans, melatonin secretion increases as light levels drop in the evening and levels peak over the nighttime hours. The target organ for melatonin appears to be the reticular formation of the brainstem, where nerve cell groups that regulate the sleep-wake cycle and the induction of the different stages of sleep are located.

The actions of the pineal gland also influence the hormonal secretion functions of the hypothalamus because many hormones released from the anterior pituitary, e.g., adrenocorticotropic hormone (ACTH) and growth hormone (GH), follow daily rhythms. Interfering with the normal light/dark cycle, e.g., by watching a bright screen late into the night, can have adverse effects on sleep and wakefulness.

▶ **REGULATION OF DAILY CYCLES BY HORMONES**

Many functions of the brain and endocrine system are regulated to follow a diurnal or daily cycle by the secretion of melatonin from the pineal gland above the thalamus of the brain. Both growth hormone and corticosteroids follow regular daily cycles of peaks and troughs of secretion. Melatonin secretion is itself regulated by light input from the retina to the hypothalamus. Note: DHEA = dehydroepiandrosterone; IGF-1 = insulin-like growth factor 1.

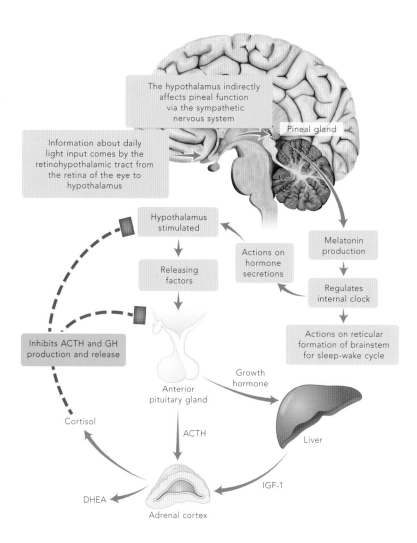

The hypothalamus indirectly affects pineal function via the sympathetic nervous system

Information about daily light input comes by the retinohypothalamic tract from the retina of the eye to hypothalamus

Pineal gland

Hypothalamus stimulated

Releasing factors

Actions on hormone secretions

Melatonin production

Regulates internal clock

Actions on reticular formation of brainstem for sleep-wake cycle

Inhibits ACTH and GH production and release

Growth hormone

Anterior pituitary gland

Cortisol

ACTH

Liver

DHEA

IGF-1

Adrenal cortex

Thyroid gland

THE THYROID GLAND IS A DOUBLE-LOBED STRUCTURE IN THE LOWER NECK AROUND THE LARYNX OR VOICE BOX. The thyroid gland is made up of multiple spherical structures called thyroid follicles. Each follicle consists of a layer of cuboidal cells (follicle cells) surrounding a space that is filled with a protein-rich gelatinous substance known as colloid. Colloid is the storage site of the precursors of thyroid hormones and also contains a high concentration of iodine, which is a key ingredient of thyroid hormone.

The spaces between the follicles contain parafollicular cells that are not involved in thyroid hormone production but are key players in calcium metabolism (see pp. 300–301).

The thyroid gland is under the direction of thyroid stimulating hormone (TSH) secreted from the anterior pituitary gland. Under TSH stimulation, the thyroid gland produces two forms of thyroid hormone: thyroxine (T_4) and triiodothyronine (T_3). Both have an amino acid core, contain iodine, and have similar physiological actions, although T_3 is more active than T_4. In fact, T_4 is converted to T_3 in the target tissue. Both T_3 and T_4 are sufficiently lipid-soluble that they can diffuse into target cells and bind to intracellular receptors in the nucleus.

The role of thyroid hormones is to regulate metabolic rate and thermogenesis, to promote growth and development, and to increase sympathetic nervous system function. Thyroid hormones feed back on the hypothalamus and anterior pituitary in a negative feedback regulatory loop. This allows thyroid hormone concentration to be maintained at a steady level in the body.

▶ **MICROSTRUCTURE OF THE THYROID GLAND**

The thyroid gland consists of multiple follicles with hollow interiors that are usually filled with a protein-rich gelatinous substance called colloid. Two parathyroid glands are attached to the back of each lobe of the thyroid gland, and small parts of both glands can be seen in this microscopic view.

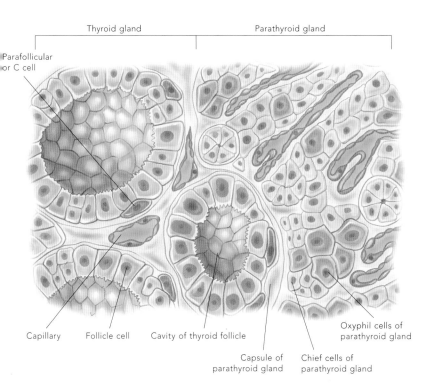

Thyroid gland | Parathyroid gland

Parafollicular or C cell

Capillary

Follicle cell

Cavity of thyroid follicle

Oxyphil cells of parathyroid gland

Capsule of parathyroid gland

Chief cells of parathyroid gland

HYPERTHYROIDISM/HYPOTHYROIDISM

Secretion of thyroid hormone can be either in excess (hyperthyroidism) or deficient (hypothyroidism). Symptoms and signs of hyperthyroidism include fast heartbeat, muscular weakness, sleeping difficulty, and poor tolerance to heat. Symptoms and signs of hypothyroidism include a feeling of tiredness, poor tolerance of cold, weight gain, and slow cognitive function. Hypothyroidism in childhood causes slow growth and intellectual disability.

Regulation of calcium metabolism

CALCIUM PLAYS A VITAL ROLE IN THE FUNCTION OF EXCITABLE TISSUES, i.e., muscles and nerves, in the body and must be maintained at a constant level (2.15 to 2.55 mmol/liter). There are two major hormones involved in calcium homeostasis. The parafollicular cells of the thyroid gland release calcitonin when blood calcium concentration is too high. Calcitonin promotes the deposition of calcium into bone by inhibiting the action of the osteoclast cells that routinely absorb bone tissue during natural bone remodeling. As calcium levels return to optimum, the secretion of calcitonin declines.

If calcium ion concentration in the blood becomes too low, the parathyroid glands release a hormone called parathyroid hormone. Parathyroid hormone promotes the mobilization of calcium from stores in the bone by stimulating the actions of the osteoclast cells that resorb bone. Parathyroid hormone also promotes reabsorption of calcium from the urine and small intestine to minimize excretory loss and increase intake from food, respectively. The action on the

▶ **CONTROL OF CALCIUM ION CONCENTRATION IN THE BLOOD**

Blood calcium concentration is regulated by two hormones: the parathyroid hormone from the parathyroid glands that raises calcium ion (Ca^{2+})concentration, and calcitonin from parafollicular or C cells of the thyroid gland that decreases calcium ion concentration.

small intestine is indirect, in that parathyroid hormone converts vitamin D to its active form, vitamin D3 or calcitriol, which in turn increases the number of calcium ions absorbed from the small intestine.

When calcium ion levels return to normal, the secretion of parathyroid hormone is gradually reduced. Calcitonin may be used as a treatment for osteoporosis, a condition in which mineral density of the bone is decreased often leading to reduced bone strength and fractures.

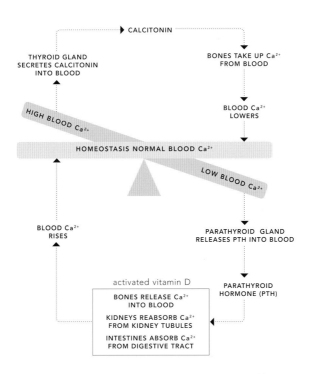

CALCITONIN

THYROID GLAND
SECRETES CALCITONIN
INTO BLOOD

BONES TAKE UP Ca^{2+}
FROM BLOOD

BLOOD Ca^{2+}
LOWERS

HIGH BLOOD Ca^{2+}

HOMEOSTASIS NORMAL BLOOD Ca^{2+}

LOW BLOOD Ca^{2+}

BLOOD Ca^{2+}
RISES

PARATHYROID GLAND
RELEASES PTH INTO BLOOD

activated vitamin D

PARATHYROID
HORMONE (PTH)

BONES RELEASE Ca^{2+}
INTO BLOOD

KIDNEYS REABSORB Ca^{2+}
FROM KIDNEY TUBULES

INTESTINES ABSORB Ca^{2+}
FROM DIGESTIVE TRACT

OSTEOPOROSIS

Osteoporosis is a condition of reduced mechanical strength of bone, leading to fractures. Osteoporosis is particularly common among women because calcium stores are used during a woman's reproductive years to meet the calcium needs of fetuses and breast-feeding neonates, leading to reduced mineral density of bone. Reduction of estrogen levels after menopause also reduces bone mineral density. Fractures can be of long bones, e.g., fractured neck of femur and distal radius, or vertebral bodies, e.g., crush fractures of thoracic vertebrae leading to the bowed back commonly known as dowager's hump. Suitable calcium intake throughout life is important to reduce the risk of developing osteoporosis.

Endocrine pancreas: cell types & functions

THE ENDOCRINE PART OF THE PANCREAS CONSISTS OF THE ISLETS OF LANGERHANS. These are about one million spherical clumps of cells distributed throughout the exocrine part of the pancreas.

Pancreatic islets secrete hormones into the bloodstream, and there are four types of cells within them. Alpha cells make up 20% of the cells in the islets. They are mainly located at the periphery of the islet and secrete a peptide hormone known as glucagon. Beta cells make up around 68% of the cells in the islets and are spread throughout the islet. Beta cells secrete a protein hormone called insulin.

Delta cells make up 10% of the cells in the islet and secrete the peptide hormones gastrin and somatostatin. Gastrin increases stomach acid secretion, while somatostatin is a polypeptide that spreads by local diffusion through the islet to inhibit the release of both insulin and glucagon. Finally, F cells make up only 2% of the total cells in the islets. They produce a chemical called pancreatic polypeptide that inhibits the secretion of somatostatin.

The pancreatic islets have a rich vascular supply to carry blood to the islets for the sensing of glucose concentration in the blood and to transport the hormones away to body tissues.

▶ **CELLS OF THE ENDOCRINE PANCREAS**

The endocrine part of the pancreas consists of the islets of Langerhans, which are about one million globules of endocrine cells distributed evenly throughout the pancreas. Each islet is richly supplied with blood so that the endocrine cells can detect blood glucose concentration and secrete their hormones into draining venules (insuloacinar portal veins). Some islet cells act on the gut, e.g., delta cells secreting gastrin and somatostatin, whereas others, e.g., alpha and beta cells secreting glucagon and insulin, exert their main effect on distant tissues.

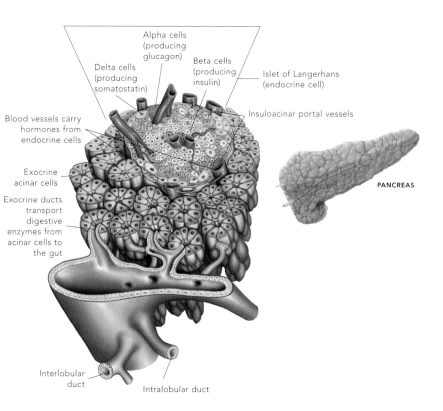

Alpha cells (producing glucagon)

Delta cells (producing somatostatin)

Beta cells (producing insulin)

Islet of Langerhans (endocrine cell)

Blood vessels carry hormones from endocrine cells

Insuloacinar portal vessels

Exocrine acinar cells

Exocrine ducts transport digestive enzymes from acinar cells to the gut

Interlobular duct

Intralobular duct

PANCREAS

Endocrine pancreas: insulin & glucagon

THE TWO MAIN HORMONES RELEASED BY THE PANCREATIC ISLET CELLS ARE BOTH CONCERNED WITH REGULATION OF BLOOD GLUCOSE CONCENTRATION. Glucagon acts on the liver, muscle, and adipose tissue. Its role is to promote processes that increase the level of glucose and other metabolic fuels in the bloodstream. It does this by promoting the breakdown of glycogen stores in the liver and muscle into glucose (glycogenolysis) and by promoting the formation of new glucose in the liver (gluconeogenesis). It can also stimulate protein breakdown to release amino acids for glucose production and the release of fats from adipose tissue for gluconeogenesis. In starvation conditions, it can also stimulate the formation of ketone bodies from fatty acids.

Insulin's effects tend to decrease blood glucose concentration. It does this by stimulating the uptake of lipids, amino acids, and glucose by the body's cells, promoting the synthesis of glycogen in the liver, stimulating the synthesis of fats from fatty acids and carbohydrates, and acting on the brain to provide a feeling of satiety so that nutrient intake is reduced. The presence of insulin is an absolute requirement for glucose uptake by almost all the cells of the body—the energy-hungry brain being the only exception. All other cells will starve without insulin, even if they are bathed in glucose. If insulin levels are too high, the blood glucose level can drop precipitously, causing severe hypoglycemia and resulting in seizures, coma, or death.

▶ **HORMONAL REGULATION OF BLOOD GLUCOSE LEVELS**
Insulin is released from beta cells of the islets of Langerhans in response to a glucose-containing meal, whereas the same meal leads to a drop in glucagon secretion from the alpha cells. Glucagon release occurs in response to low blood glucose concentration.

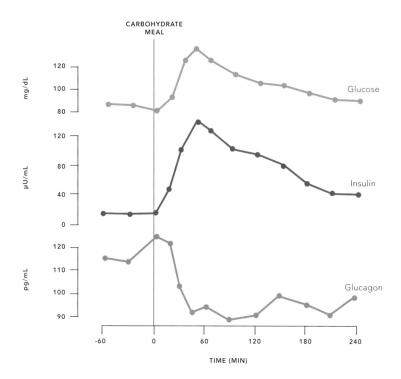

Adrenal cortex structure & function: aldosterone

THE ADRENAL CORTEX IS THE OUTER PART OF THE ADRENAL GLAND. The adrenal cortex is divided into three zones: an outer zona glomerulosa with densely packed cells, a middle zona fasciculata with columns of cells, and an inner zona reticularis with loose clumps of cells.

The zona glomerulosa produces a group of steroid hormones called mineralocorticoids. Mineralocorticoids regulate the concentration of sodium and potassium, and the main hormone of the group is aldosterone. The functions of aldosterone include: maintenance of extracellular sodium and potassium ions within their optimal ranges, regulation of extracellular fluid volume, maintenance of blood pressure, and maintenance of acid-base balance.

Aldosterone acts on the kidney tubules to stimulate reabsorption of sodium and chloride from the kidney tubules into the extracellular space and secretion of potassium into the kidney tubules. The actions of aldosterone on sodium and chloride create a concentration gradient in the kidney that promotes the osmosis of water from the kidney filtrate into the extracellular space. Blood volume is a major determinant of blood pressure, so fluid retention caused by aldosterone also gives rise to increased blood pressure.

Several factors can stimulate the secretion of aldosterone, but the main causes are elevated blood potassium ion concentration, a decrease in blood pH, and the effects of a hormone called angiotensin II. Adrenocorticotropic hormone (ACTH) can also stimulate the production of aldosterone.

▶ **STRUCTURE AND FUNCTION OF THE ADRENAL CORTEX**

The adrenal glands are situated on the upper poles of each kidney. Each adrenal gland is divided into an outer cortex and an inner medulla. The outermost part of the adrenal cortex (the zona glomerulosa) makes the hormone aldosterone, which regulates the concentration of sodium and potassium in the blood.

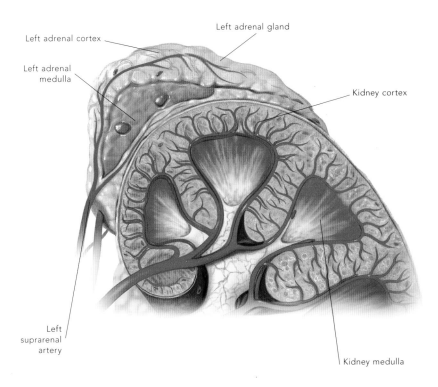

Left adrenal gland

Left adrenal cortex

Left adrenal
medulla

Kidney cortex

Left
suprarenal
artery

Kidney medulla

Adrenal cortex structure & function: cortisol

GLUCOCORTICOIDS ARE STEROID HORMONES PRODUCED IN THE ZONA FASCICULATA AND ZONA RETICULARIS OF THE ADRENAL CORTEX. The main role of glucocorticoids is to help the body respond to stress, and, as their name implies, they have an effect on blood glucose levels.

The most powerful of the glucocorticoids is cortisol or hydrocortisone. Cortisol mainly acts on liver muscle and adipose tissue. Its actions on the liver are to promote gluconeogenesis, a process in which amino acids and fats are converted to glucose, the effect being to raise blood glucose levels. The effect of cortisol on skeletal muscle is to induce the breakdown of muscle proteins to release amino acids to make glucose by gluconeogenesis in the liver. Cortisol also causes the release of fatty acids from adipose tissue. Fatty acids can be used as an alternative source of energy to glucose and by the liver in gluconeogenesis.

Most tissues in the body have receptors for cortisol, so its effects are widespread. A major effect of cortisol is on the immune system, where it acts as an anti-inflammatory agent by lowering the populations of white blood cells in the blood. It is for this reason that corticosteroids are used as anti-inflammatory agents in individuals with autoimmune disease and in transplant recipients.

Cortisol secretion is under the control of adrenocorticotropic hormone (ACTH) from the anterior pituitary, and cortisol feeds back onto the hypothalamus and anterior pituitary to lower ACTH synthesis. This negative feedback regulatory loop keeps cortisol levels at optimal concentrations.

▶ **MICROSTRUCTURE OF THE ADRENAL CORTEX**

The zona fasciculata and reticularis of the adrenal cortex make glucocorticoids. The effects of glucocorticoids are predominantly a stress response, raising blood glucose levels to deal with an emergency and toning down the inflammatory response.

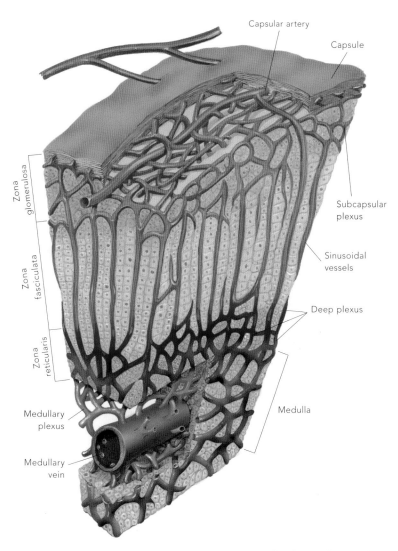

Capsular artery

Capsule

Zona glomerulosa

Zona fasciculata

Zona reticularis

Subcapsular plexus

Sinusoidal vessels

Deep plexus

Medullary plexus

Medullary vein

Medulla

Adrenal cortex structure & function: sex hormones

THE GONADS (OVARIES AND TESTES) MAKE MOST SEX STEROIDS, but androgenic (literally "male-producing") steroids are made in small quantities in the adrenal cortex in both males and females. This is mainly a by-product of the production of cortisol, but the adrenal androgens can still have significant effects on the body and play a major role in sexual development and libido.

Genetic defects in some of the enzymes that produce cortisol and mineralocorticoids, e.g., 21-hydroxylase, can lead to such large amounts of androgenic steroids being made by the adrenal cortex that precocious sexual development occurs in the affected children (congenital adrenogenital syndrome). The clitoris of affected girls can grow so large that it looks like a penis and the child can be assigned masculine gender.

Affected children have symptoms from both inadequate production of mineralocorticoids, e.g., vomiting, salt-wasting, and dehydration, and from excess production of adrenal androgens, e.g., virilization, facial hair growth, precocious puberty, and infertility. Inefficient cortisol production is less severe in the affected children than the deficiency of mineralocorticoids, which can be fatal.

▶ **SEX HORMONE PRODUCTION BY THE ADRENAL CORTEX**

In addition to glucocorticoids and mineralocorticoids, the adrenal cortex has the capacity to produce sex steroids like DHEA (dehydroepiandrosterone) and androstenedione. These are important during fetal development and when the gonadal production of sex steroids declines, e.g., after menopause. They also have a significant effect if there is a genetic deficiency in one of the enzymes that produces glucocorticoids or mineralocorticoids.

STEROID HORMONE SYNTHESIS PATHWAYS

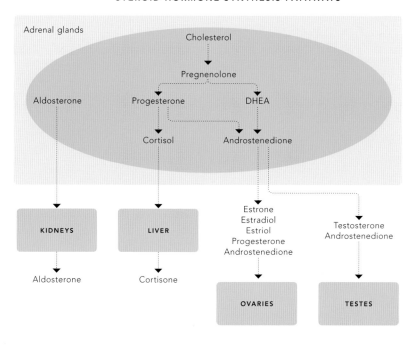

Adrenal medulla & catecholamines

THE ADRENAL MEDULLA IS THE CENTRAL PART OF THE ADRENAL GLAND. In fact, the cells of this region are modified nerve cells called chromaffin cells. These cells are stimulated by acetylcholine released from preganglionic nerve cells of the sympathetic nervous system. The chromaffin cells are very similar to the postganglionic nerve cells of the sympathetic nervous system, except that they secrete their products directly into the bloodstream. They produce chemicals called catecholamines (both epinephrine and norepinephrine, but mainly epinephrine) that can diffuse through the body.

Catecholamines from the adrenal medulla produce the body's immediate response to a danger or stressor by a variety of effects. These include: increasing the force and rate of cardiac contractions so that the heart can more effectively pump blood around the body; dilating the small airways (bronchioles) to improve lung ventilation and bring more oxygen into the lungs; constricting the blood vessels that supply the skin, digestive system, and urinary organs so that oxygen and nutrient-rich blood can be available for the skeletal muscle in an emergency; increasing the blood flow to skeletal muscle so that escape from danger is more effective; dilating the pupils to increase light reaching the retina; and decreasing the function of the digestive and urinary systems so that energy is diverted to skeletal muscles.

Adrenocorticotropic hormone (ACTH) and cortisol can also stimulate adrenal medullary secretion as part of a generic stress response.

▶ **THE ADRENAL MEDULLA AND CATECHOLAMINES**

The hormones of the adrenal medulla are secreted in response to potentially threatening circumstances and are made from the amino acid tyrosine by adding hydroxyl and methyl groups. The effects of adrenal medullary hormones are to improve the emergency performance of muscles (both cardiac and skeletal) so that danger can be escaped. They also make more energy available by breaking down carbohydrate reserves in the liver to raise glucose concentration in the blood.

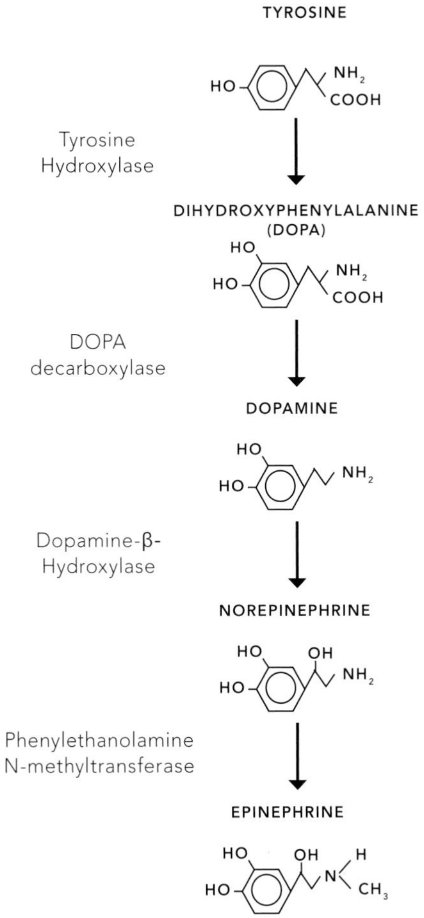
TYROSINE
HO
NH2
COOH
Tyrosine
Hydroxylase
DIHYDROXYPHENYLALANINE
(DOPA)
HO
HO
NH2
COOH
DOPA
decarboxylase
DOPAMINE
HO
HO
NH2
Dopamine-β-
Hydroxylase
NOREPINEPHRINE
HO
OH
HO
NH2
Phenylethanolamine
N-methyltransferase
EPINEPHRINE
HO
OH
H
N
HO
CH3

Gonadal hormones

APART FROM PRODUCING THE SEX CELLS (sperm and ova), the gonads (testes and ovaries) are important sites of hormone production. The testes produce the steroid hormone testosterone, which has two basic effects: anabolic effects, i.e., stimulation of bone growth and increase in muscle mass, and androgenic effects, i.e., the development of male secondary sexual characteristics such as a deeper voice and prominent facial hair.

Testosterone production is regulated by gonadotrophin-releasing hormone (GnRH) from the hypothalamus and luteinizing hormone (LH) and follicle-stimulating hormone (FSH) from the anterior pituitary gland. LH stimulates testosterone production itself, whereas FSH stimulates production of a protein that binds and concentrates testosterone. The production of testosterone is under negative feedback regulation, with testosterone feeding back to reduce GnRH, FSH, and LH production.

The ovary produces estrogen under direction of GnRH from the hypothalamus, and FSH and LH from the anterior pituitary, with estrogen acting on the sites of regulatory hormone production by negative feedback. Estrogen produces female characteristics, e.g., breast development and feminine fat distribution, and regulates the menstrual cycle. The ovaries also produce progesterone, which peaks in concentration in the menstrual cycle after ovulation and continues to be produced during pregnancy. Progesterone has multiple effects, including preparing the body for pregnancy and supporting fetal development. It may also affect the smooth muscle tissue of the uterus, body temperature, and the tendency of the blood to clot.

▶ **TESTES, OVARIES, AND SEX HORMONES**

The gonads (testes in males, ovaries in females) produce sex hormones in response to anterior pituitary hormones (FSH, LH) and gonadotrophin-releasing hormone (GnRH) from the hypothalamus. The sex hormones (testosterone in males and estrogens in females) are responsible for the secondary sexual characteristics of each gender.

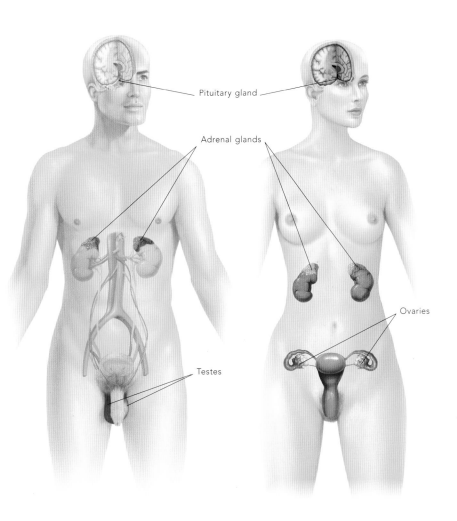

Pituitary gland

Adrenal glands

Ovaries

Testes

The effects of prostaglandins on the endocrine system

PROSTAGLANDINS BELONG TO A GROUP OF MOLECULES CALLED THE EICOSANOIDS AND LEUKOTRIENES. These are lipid-based, cell-signaling molecules that can bind to cell surface receptors (unlike the steroid hormones, which must enter the cell). The eicosanoid group of molecules is important in a diverse range of functions, including stimulating the clumping of platelets during hemostasis, creating inflammatory responses to injury, and contracting smooth muscle in the uterus and fetal blood vessels around the time of birth.

The eicosanoid group of compounds is synthesized from arachidonic acid, which is converted to prostaglandin H_2 by an enzyme called prostaglandin synthase. This is an important synthetic step because prostaglandin synthase can be inhibited by aspirin and other nonsteroidal anti-inflammatory drugs. The effects of aspirin—lessening of pain, reduction of inflammation, inhibition of the aggregation of platelets, and reduction of the coagulation of the blood—are all due to the inhibition of prostaglandin synthase. The effects on platelets and blood coagulation make aspirin an effective prophylaxis against strokes.

▶ **THE ACTIONS OF PROSTAGLANDINS**

Prostaglandins have diverse effects throughout the body. The illustration here shows the involvement of prostaglandins and other cell signals in the response of the tissues to injury. These factors produce the key components of inflammation: swelling, pain, and heat and redness (from vascular dilation). Prostaglandins also play key roles in the contraction of vascular smooth muscle and in platelet aggregation.

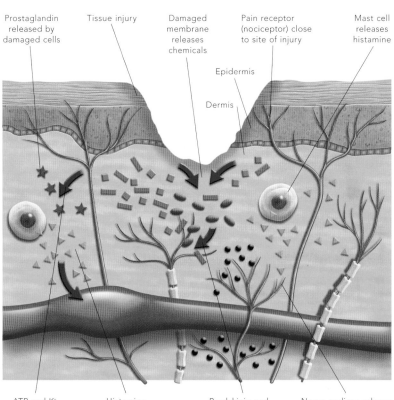

Prostaglandin released by damaged cells

Tissue injury

Damaged membrane releases chemicals

Pain receptor (nociceptor) close to site of injury

Mast cell releases histamine

Epidermis

Dermis

ATP and K⁺ break down to form bradykinin

Histamine causes capillary to swell

Bradykinin and ATP bind to nerve receptors

Nerve endings release substance P, which stimulates other nerve cells to do the same

★ PROSTAGLANDIN

▲ HISTAMINE

● SUBSTANCE P

▮ K⁺ (POTASSIUM)

〰 BRADYKININ

▦ ATP (ADENOSINE TRIPHOSPHATE)

Overview of structure of the female reproductive system

THE FEMALE REPRODUCTIVE SYSTEM CONSISTS OF paired ovaries for the production of eggs (oocytes) and for hormonal regulation of the menstrual cycle and pregnancy; uterine tubes that carry sperm cells toward the oocyte and the newly fertilized zygote down toward the uterus; a uterus for the new embryo to implant and develop into a fetus; a vagina to receive the male penis during sexual intercourse and to act as a birth canal; and external genitalia, which protect the entrance to the vagina and provide lubrication during intercourse.

From the age of menarche (first menstrual cycle) to menopause (the last menstrual cycle), the female reproductive system undergoes rhythmic changes every 28 days called the menstrual cycle. This cycle involves hormonal fluctuations that stimulate the ovaries to produce and release an oocyte in the middle of the cycle (ovulation) and prepare the reproductive tract for possible fertilization and implantation in the uterine wall.

The uterus undergoes significant changes if an embryo implants in its wall, participating in the development of a placenta to nourish the embryo and fetus and becoming thicker with smooth muscle in preparation for the process of labor, when the full-term fetus is expelled down the birth canal by muscular contractions of the uterine smooth muscle.

▶ **COMPONENTS OF THE FEMALE REPRODUCTIVE SYSTEM**

The female reproductive system consists of paired ovaries to produce oocytes and female sex hormones, uterine or Fallopian tubes to convey oocytes to the uterus for implantation and gestation of the embryo, and a vagina to receive semen and provide a birth canal for the fetus at the end of gestation.

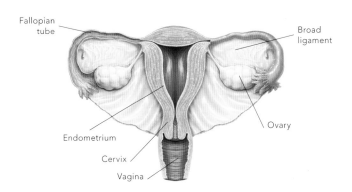

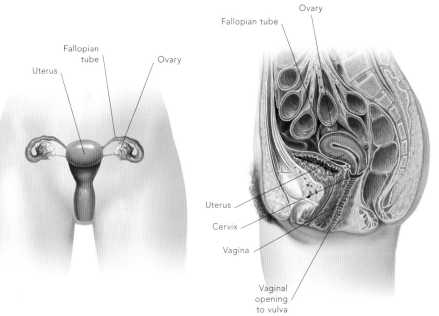

Ovary: formation of the oocyte & ovulation

DURING THE MENSTRUAL CYCLE THE OVARY GOES THROUGH THREE PHASES: a follicular phase, an ovulatory phase, and a luteal phase. The follicular phase is the stage during which a primordial follicle develops into a mature follicle under the influence of follicle-stimulating hormone (FSH) from the anterior pituitary. At birth, each ovary contains two to four million primordial follicles, each of which contains a primary oocyte (immature egg cell). During the follicular stage, the follicular cells around the oocyte become cuboidal to form granulosa cells. Microvilli develop in the oocyte and grow into the surrounding granulosa cells to form a glycoprotein-rich zone around the oocyte called the zona pellucida. The microvilli of the oocyte increase the surface area to allow the granulosa cells to transfer nutrients to the oocyte.

During the secondary follicle stage, the granulosa cells begin to secrete follicular fluid into the space around the oocyte. When these pockets of fluid coalesce, they form a single large cavity or antrum surrounded by follicular cells and theca interna and externa cell layers.

At this point the follicle is a vesicular or tertiary follicle (mature Graafian follicle). Ovulation is the process by which the ovary expels the oocyte and its granulosa cells. This occurs at the end of the follicular phase, when surges in FSH and luteinizing hormone production from the anterior pituitary occur.

After the oocyte and surrounding corona radiata of granulosa cells have been expelled into the abdominal cavity near the opening of the uterine tube, the follicle proceeds to a luteal phase, when the remnants of the ruptured follicle become a corpus luteum that will produce progesterone to help maintain any pregnancy that might occur.

▶ **STRUCTURE AND FUNCTION OF THE OVARY**

The paired ovaries are responsible for the production and release of the oocyte (egg or ovum); for the production of estrogen to induce secondary sexual characteristics; and for the production of estrogen and progesterone to induce proliferation, secretion, and support of the uterine lining. They lie alongside the lateral pelvic wall adjacent to the opening of the Fallopian tube between the layers of the broad ligament.

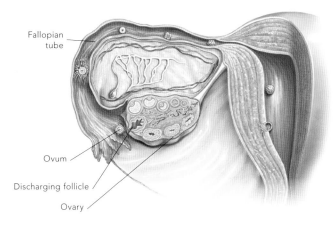

Fallopian tube

Ovum

Discharging follicle

Ovary

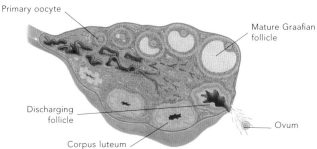

Primary oocyte

Mature Graafian follicle

Discharging follicle

Corpus luteum

Ovum

POLYCYSTIC OVARY SYNDROME

Polycystic ovary syndrome is a common condition (5% to 10% of women of reproductive years) in which there is excessive production of male hormones (androgens). The condition is characterized by irregular or absent menstruation, infertility, excess facial hair, and acne. Risk factors include obesity, family history, and lack of physical activity. Treatment is primarily by changing lifestyle, i.e., weight loss and increased exercise, and prescription of the contraceptive pill to make menstrual cycles more regular.

Ovary: hormone production

OVARIAN FUNCTION IS REGULATED BY A RHYTHMIC CYCLE OF HORMONE PRODUCTION BY THE HYPOTHALAMUS, the anterior pituitary, and the ovaries themselves. The hypothalamus produces gonadotrophin-releasing hormone (GnRH) that acts on the anterior pituitary to produce follicle-stimulating hormone (FSH) and luteinizing hormone (LH).

FSH stimulates the development of the follicle, while LH stimulates the thecal cells around the follicle to produce androgens, which diffuse locally and act on the granulosa cells. FSH also stimulates the granulosa cells to produce estrogens and to convert the androgens from the surrounding cells to produce even more estrogens.

The estrogens made by the granulosa cells stimulate the secondary follicle to develop into a mature vesicular Graafian follicle ready for rupture. When the vesicular follicle is big enough, it produces enough estrogens to exert a positive feedback on the anterior pituitary to induce the surge of LH that triggers ovulation.

Once the oocyte has been expelled, the corpus luteum produces progesterone, estrogen, and a hormone called inhibin. Inhibin has a negative feedback effect on the anterior pituitary to decrease FSH secretion, and this inhibits the maturation of other follicles. If fertilization occurs, the progesterone from the corpus luteum will inhibit the release of any further gonadotrophins from the anterior pituitary. If fertilization doesn't occur, the corpus luteum degenerates to a corpus albicans and the anterior pituitary is released from any inhibition so the cycle can begin anew.

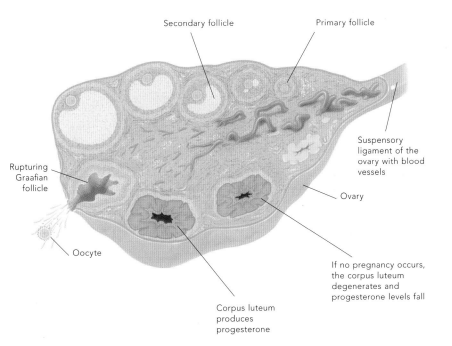

Secondary follicle

Primary follicle

Suspensory ligament of the ovary with blood vessels

Rupturing Graafian follicle

Ovary

Oocyte

If no pregnancy occurs, the corpus luteum degenerates and progesterone levels fall

Corpus luteum produces progesterone

▲ CYCLICAL CHANGES IN THE OVARY

The cyclical changes in the ovary are represented here in a series of illustrations of the developing ovarian follicle. In the early part of the menstrual cycle, FSH from the anterior pituitary stimulates the formation of an ovarian follicle, while LH stimulates the thecal cells around the follicle to produce androgens. FSH also stimulates the granulosa cells around the follicle to make estrogen. At day 14 of the cycle, the follicle ruptures, releasing the oocyte, and the remains of the follicle turn into a corpus luteum, which produces progesterone and estrogen to prepare the uterine wall for possible implantation of a fertilized oocyte.

Structure & function of the uterine wall

THE ROLE OF THE UTERUS IS TO ACCEPT A SIX-DAY EMBRYO, allow it to implant in the uterine wall and develop into an embryo and then fetus, and finally deliver the fetus at the end of pregnancy. The wall of the uterus is divided into three layers. The inner lining is called the endometrium, and this layer undergoes cyclical changes during the menstrual cycle in response to hormones secreted from the ovary.

During days 6 to 14 of each menstrual cycle, the endometrium becomes thicker, more glandular, and more vascular (proliferative phase). After ovulation (days 15 to 28 of the menstrual cycle), the glands of the uterine wall secrete a glycogen-rich fluid known as uterine milk into the uterine cavity to sustain the new embryo. If fertilization has not occurred or the embryo fails to implant, the endometrium is sloughed off in a menstrual phase (days 1 to 6 of the menstrual cycle), and the cycle begins anew.

The uterus has a thick layer of smooth muscle called the myometrium outside the endometrium. The myometrium contracts during sexual intercourse to assist movement of sperm and also during expulsion of the sloughed endometrium during menstruation, but its main role is to expel the fetus during parturition (childbirth). Special pacemaker cells found close to the uterine horns at the top of the uterine body set a rhythmic activation that spreads through the uterine muscle wall so that labor contractions are coordinated and rhythmic.

▶ **STRUCTURE OF THE UTERUS**

The uterine wall is divided into three layers: a glandular endometrium, a smooth muscle myometrium, and a connective tissue adventitia or serosa. The uterus receives the openings of the Fallopian tubes that carry the oocyte down from the ovary. The uterus also opens into the vagina to allow sperm to enter the uterine cavity and to expel the fetus at the end of gestation. The uterus has a fundus above, a central body, and a cervix with internal and external openings (internal os and external os, respectively). The uterus is enveloped in a double-layered membrane called the broad ligament that has three regions: the mesovarium near the ovary, the mesosalpinx attached to the uterine tube, and the mesometrium alongside the uterine body.

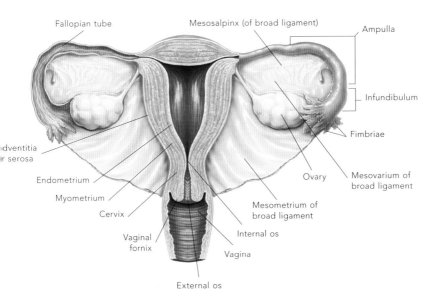

Fallopian tube

Mesosalpinx (of broad ligament)

Ampulla

Infundibulum

Fimbriae

Adventitia or serosa

Endometrium

Myometrium

Cervix

Vaginal fornix

External os

Ovary

Mesovarium of broad ligament

Mesometrium of broad ligament

Internal os

Vagina

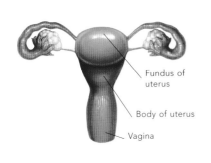

Fundus of uterus

Body of uterus

Vagina

ENDOMETRIOSIS

Endometriosis is a condition in which small fragments of uterine endometrium become lodged in the pelvic cavity. A cause of this may be prior pelvic inflammatory disease. The pieces of ectopic endometrium undergo changes with the menstrual cycle, thickening in the first half and degenerating at the end of the cycle. Degeneration and bleeding of the ectopic endometrium each month causes pelvic pain and scarring and may lead to infertility.

Hormonal regulation of the menstrual cycle

THE MENSTRUAL CYCLE LASTS ON AVERAGE 28 DAYS. It is under the regulation of cyclical changes in hormones secreted from the hypothalamus, anterior pituitary, and ovaries. Days 1 to 6 of the menstrual cycle are said to be the menstrual phase of the uterus, when the endometrium built up during the previous cycle is sloughed off and expelled from the cervix. During this phase, the levels of follicle-stimulating hormone (FSH) and luteinizing hormone (LH) begin to rise, but it is not until the end of the menstrual phase that the ovarian follicles start to increase the production of estrogen.

From days 7 to 13 (proliferative phase of the uterus), the estrogen levels rise steadily, whereas progesterone levels stay low. This leads to thickening of the endometrium and elongation of the endometrium glands and blood vessels. The menstrual and proliferative phases of the uterus together correspond to the follicular phase of the ovary. Leading up to day 14, LH and FSH production by the anterior pituitary rise rapidly, triggering ovulation.

From days 15 to 28 (the secretory phase of the uterus and the luteal phase of the ovary), progesterone production by the corpus luteum rises and plateaus, and FSH and LH production drops due to negative feedback on the anterior pituitary. If fertilization does not occur, progesterone production by the corpus luteum begins to drop at around day 24. The endometrium begins to break down and slough away at day 28, beginning the cycle anew.

▶ **CYCLICAL CHANGES IN THE ENDOMETRIUM**

The menstrual cycle is defined as beginning with menstruation (top left), followed by a proliferative phase, when endometrial glands lengthen (top right); ovulation (release of the ovum from the ovary) at day 14 (bottom right); and a secretory phase, when the uterus prepares to receive a fertilized ovum (bottom left). Cyclical changes in blood levels of pituitary and ovarian hormones also mark the stages.

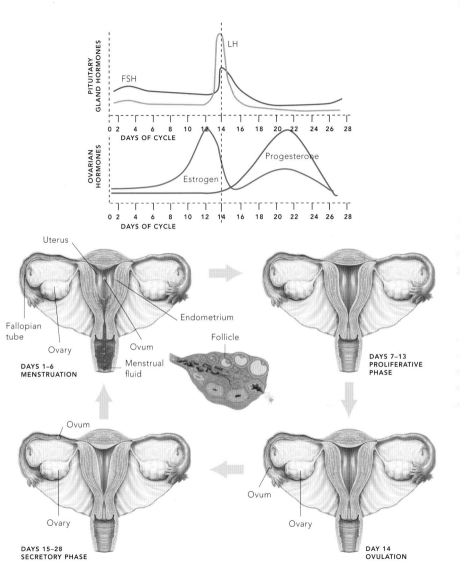

PITUITARY GLAND HORMONES

FSH

LH

DAYS OF CYCLE
0 2 4 6 8 10 12 14 16 18 20 22 24 26 28

OVARIAN HORMONES

Estrogen

Progesterone

DAYS OF CYCLE
0 2 4 6 8 10 12 14 16 18 20 22 24 26 28

Uterus

Endometrium

Fallopian tube

Ovary

Ovum

Menstrual fluid

DAYS 1–6
MENSTRUATION

Follicle

DAYS 7–13
PROLIFERATIVE PHASE

Ovum

Ovary

DAYS 15–28
SECRETORY PHASE

Ovum

Ovary

DAY 14
OVULATION

Physiology of the human female sexual response

THE ROLE OF THE FEMALE SEXUAL RESPONSE IS TO PROVIDE LUBRICATION FOR THE PENETRATION OF THE VAGINA BY THE PENIS. It also assists the movement of sperm cells up the female reproductive tract toward the oocyte. During sexual arousal, the vaginal mucosa, erectile tissue of the vaginal entrance, and breasts become engorged with blood. The vagina produces a lubricating fluid by the transudation or leakage of fluid across the walls of the capillaries of the engorged mucosa. The greater vestibular glands at the entrance of the vagina also produce copious amounts of lubrication for the vaginal entrance. Changes also occur in the nipples, which become erect, and the skin of the face and chest, which become flushed. Heart and breathing rate increase and the pupils dilate.

When the woman reaches orgasm, the smooth muscle of the uterus begins to have rhythmic contractions and the cervix of the uterus pushes down into the vagina. These contractions are accompanied by intense pleasure. If semen has been deposited in the vagina around the cervix, the contractions of the cervix push the cervical canal opening into the pool of semen, and the rhythmic contractions draw sperm cells into the uterine cavity and help propel them toward the oocyte. Women have no refractory period in their sexual response, so they can experience multiple orgasms during one sexual encounter. Nevertheless, orgasm is not essential for conception to occur.

Female libido or sex drive is reliant on circulating levels of androgens, e.g., dehydroepiandrosterone, produced in the adrenal cortex, but estrogens are also important. Libido often fluctuates during the menstrual cycle, usually peaking around the time of ovulation.

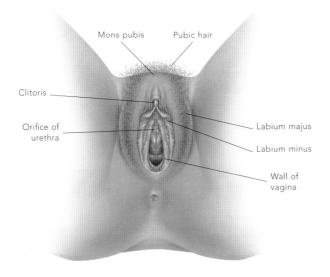

Mons pubis   Pubic hair

Clitoris

Orifice of
urethra

Labium majus

Labium minus

Wall of
vagina

▲ THE FEMALE EXTERNAL GENITALIA

The entrance to the vagina is flanked by folds of thin skin
called the labia minora (singular—labium minus) and fleshy
external folds called the labia majora (singular—labium
majus). The vaginal entrance is directly behind the orifice of
the urethra, so the urethra can be abraded during vigorous
sexual intercourse, contributing to the risk of urinary tract
infection. The front of the genital area has a fleshy mound
called the mons pubis. The labia majora and the mons pubis
are covered in pubic hair after puberty. The clitoris is an
erectile and sensory organ that becomes erect during arousal
and is stimulated by traction on the labia during intercourse,
contributing to the female orgasm.

Menopause

MENOPAUSE LITERALLY MEANS A STOP IN MENSTRUATION, but the changes that occur in a woman's body around the age of 50 years are more accurately called the climacteric. These changes are predominantly due to the loss of effective estrogen support of key tissues in the body, e.g., genitourinary tract, breast, and brain, as estrogen production by the ovaries declines.

The climacteric takes place over several years and is divided into three phases. The premenopausal phase is a period when menstrual periods become irregular and heavy. In the perimenopausal period (a time around the actual cessation of periods), the woman will experience several symptoms, such as hot flashes or flushes (a feeling of sudden heat), night sweats, palpitations, sleep problems, depression and anxiety, and loss of libido.

The postmenopausal period follows and is characterized by several long-term effects from the loss of estrogen. The most important of these is the loss of bone density (osteoporosis), which can have serious

consequences in the form of increased susceptibility to fractures—wrist, hip, and thoracic vertebral bodies, in particular. Hip fractures can cause immobility and increased susceptibility to chest infections. Vertebral fractures can change the shape of the chest cavity and interfere with lung ventilation. Other effects include increased susceptibility to vaginal yeast infections (thrush), problems with bladder and urinary sphincter control, dry skin with wrinkling, masculine-like hair on the face, loss of muscle strength, increased risk of cardiac disease, and flattening of the breasts. The effects of menopause can be treated by hormonal supplementation (hormone replacement therapy) for a period of several years.

▶ **EFFECTS OF ESTROGEN LOSS**

Withdrawal of the hormone estrogen such as occurs at menopause has significant effects on bone mineral density leading to osteoporosis (top right) with an increased risk of fracture of the wrist, femur, vertebrae, collar bone (clavicle), and humerus. Fat distribution also shifts to the abdomen and accumulates around internal organs and under abdominal skin.

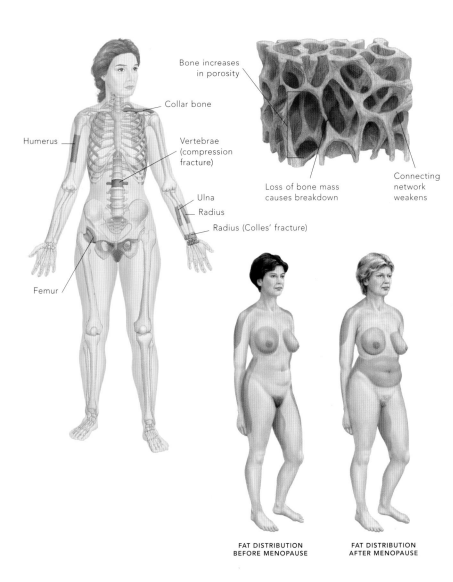

Bone increases in porosity

Collar bone

Humerus

Vertebrae (compression fracture)

Ulna

Radius

Radius (Colles' fracture)

Femur

Loss of bone mass causes breakdown

Connecting network weakens

FAT DISTRIBUTION BEFORE MENOPAUSE

FAT DISTRIBUTION AFTER MENOPAUSE

Overview of the male reproductive system

THE MALE REPRODUCTIVE SYSTEM CONSISTS OF THE TESTES, epididymis, ductus deferens, accessory glands, and the penis. The paired testes produce sperm cells (spermatozoa) and the male steroid hormone testosterone that produces male secondary sexual characteristics. The paired epididymises are arranged around the upper poles and posterior borders of the testes and assist sperm maturation. The two ductus deferens (also called vas deferens) carry sperm from each epididymis upward to the anterior abdominal wall before passing through the wall and ending at the male accessory glands (seminal vesicles and prostate) at the base of the urinary bladder. The seminal vesicles, prostate, and bulbourethral glands produce the various constituents of the semen that will be ejaculated, including fructose for sperm energy and factors that stimulate uterine contraction in the female.

The male erectile organ (the penis) consists of three bodies of erectile tissue—paired corpora cavernosa, which provide most of the stiffness of the erect penis, and the single midline corpus spongiosum, through which the urethra passes. The erectile tissue can be filled with blood under pressure to become so rigid that the penis can penetrate the female vagina. This allows the penis to deposit sperm cells along with support secretions against the cervix of the uterus so that sperm have the optimal chance of surviving to fertilize the oocyte.

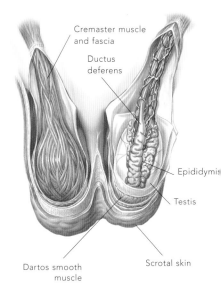

Cremaster muscle and fascia

Ductus deferens

Epididymis

Testis

Scrotal skin

Dartos smooth muscle

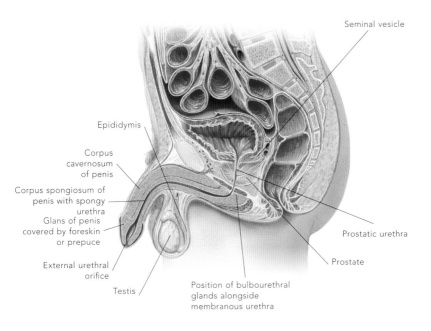

Seminal vesicle

Epididymis

Corpus cavernosum of penis

Corpus spongiosum of penis with spongy urethra

Glans of penis covered by foreskin or prepuce

External urethral orifice

Prostatic urethra

Prostate

Testis

Position of bulbourethral glands alongside membranous urethra

▲ COMPONENTS OF THE MALE REPRODUCTIVE SYSTEM

The male reproductive system consists of paired testes (singular—testis), which are held in a scrotal sac that has a smooth muscle wall (the dartos). The testes must be kept at an optimal temperature for sperm production. In cold weather the dartos and cremaster muscle contract to raise the testes closer to the body. In hot weather they relax to let the testes drop away from the body. Sperm pass through the epididymis and ductus deferens to enter the abdominal cavity and join with the ducts of the prostate and paired seminal vesicle glands at the base of the urinary bladder. Sperm and accessory gland secretions are expelled along the prostatic, membranous, and spongy urethra to the external environment during ejaculation.

Components & production of semen

SEMEN IS THE FLUID EJECTED FROM THE PENIS OF A SEXUALLY MATURE MALE AT EJACULATION. It contains spermatozoa (sperm cells) produced by the testes as well as the secretions of the accessory glands of the male reproductive system, i.e., the bulbourethral glands, seminal vesicles, and prostate gland. The bulk of semen is made by the seminal vesicles (about 70% by volume), followed by the prostate gland (about 27% by volume).

The paired bulbourethral glands produce a clear mucoid fluid (less than 1% of semen) that is expelled first to lubricate the urethra and clear it of residual urine. The paired seminal vesicles produce a yellow viscous secretion rich in fructose, which is the energy source for the spermatozoa, as well as prostaglandins to prevent an immune reaction by the female reproductive tract against sperm cells. The prostate produces a milky secretion rich in enzymes to break down protein, acid phosphatase, citric acid, and zinc. The last is important in stabilizing the DNA of sperm cells.

A normal sperm cell has a head capped by the acrosome and containing the nucleus with the DNA that will fertilize the egg, followed by a neck and body where the mitochondria that will produce energy for swimming are located. The acrosome contains enzymes that break down the outer membrane of the egg, allowing the nucleus of the sperm to join with that of the egg. Most of the length of a sperm cell is taken up by a tail that propels the sperm cell through the fluid environment of the female reproductive system.

▶ STRUCTURE OF A SPERM CELL (SPERMATOZOON)

Mature sperm consist of a head, a tail, and a connecting piece. The head contains a nucleus with the genetic material from the father, surrounded by the acrosome, which contains hydrolytic enzymes, e.g., proteases, acid phosphatases, hyaluronidase, and neuraminidase, that allow the sperm to penetrate the corona radiata and zona pellucida of the ovum to achieve fertilization. The front part of the tail consists of mitochondria arranged in a helical spiral to provide the energy for movement of the tail when the sperm swims.

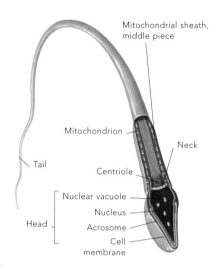

Mitochondrial sheath, middle piece

Mitochondrion

Neck

Tail

Centriole

Nuclear vacuole

Nucleus

Head

Acrosome

Cell membrane

Volume of ejaculate:	2 to 5 ml
Concentration of spermatozoa:	20 million to 100 million per ml
pH:	7.2 to 7.7
Percentage of motile forms:	more than 60%
Percentage of abnormal forms:	less than 40%
Fructose:	224 mg/dl

Spermatogenesis

SPERM CELLS ARE THE MATURE MALE REPRODUCTIVE CELLS, one of which will combine with the oocyte of the female to produce an embryo. Sperm cells are produced in the testes in a process known as spermatogenesis. The testes are suspended in the scrotal sac and supplied with blood from the testicular artery. Optimal temperature for sperm production is a few degrees below the core temperature of the body, so keeping the testes below the rest of the body is essential for fertility.

Each testicular artery is surrounded by a venous plexus (the pampiniform plexus) that helps to cool the testicular arterial blood before it reaches the testis (countercurrent heat exchange). Each testis is surrounded by a dense connective tissue sheath known as the tunica albuginea and divided into lobules by connective tissue septae. Within the lobules are up to four coiled seminiferous tubules, where sperm are produced.

Sperm are produced at the rate of about 50,000 every minute from puberty until late in life. Sperm cell production depends on rapid cell division of the germ line cells in the testis (spermatogonia and spermatocytes) to produce the spermatid stage and finally the spermatozoa. Sertoli cells (nurse cells) help in the process of spermatogenesis. The tissue around the tubules contains Leydig cells that produce testosterone.

Sperm pass through the rete testis in the mediastinum testis region and efferent ductules to reach the head of the epididymis, where they are stored for up to three months. They then pass through the tail of the epididymis and ductus deferens to be ejaculated.

▶ **STRUCTURE AND FUNCTION OF THE TESTIS AND EPIDIDYMIS**

Spermatozoa are produced in the walls of the seminiferous tubules by transformation of spermatogonia to spermatocytes, then spermatids, and finally spermatozoa. The spermatozoa then move through the rete testis (a network of tubules) and the efferent ductules to reach the tubules of the epididymis before passing along the ductus deferens to the rest of the male reproductive tract.

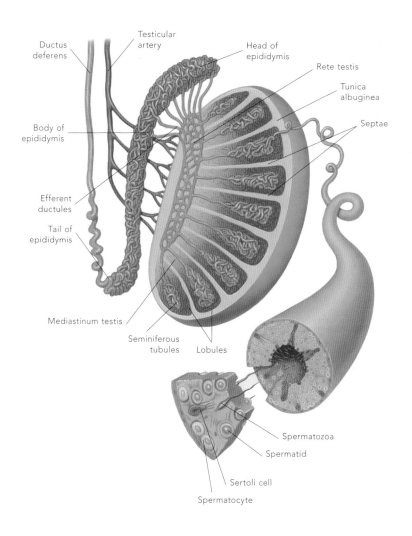

Ductus deferens

Testicular artery

Head of epididymis

Rete testis

Tunica albuginea

Septae

Body of epididymis

Efferent ductules

Tail of epididymis

Mediastinum testis

Seminiferous tubules

Lobules

Spermatozoa

Spermatid

Sertoli cell

Spermatocyte

Erection & ejaculation

ERECTION IS THE PROCESS OF FILLING CAVERNOUS SPACES WITHIN THE PENIS WITH BLOOD UNDER PRESSURE. This is necessary so that the penis can penetrate the female vagina and deposit semen as close as possible to the uterine cervix.

Erection is under control of the parasympathetic nervous system. The initial event in erection is dilation of the small arteries supplying the penis. This increases the pressure in the cavernous spaces of the paired corpora cavernosa and corpus spongiosum of the penis. The initial rise in pressure compresses those veins in the periphery of the erectile tissue that drain blood away. By a combination of increased inflow and decreased outflow, blood accumulates in the erectile tissue until the cavernous spaces are engorged with blood and the erectile tissue is rigid.

Ejaculation is under sympathetic control. When the penile shaft is stimulated, reflex centers in the sacral spinal cord initiate rhythmic contractions in the smooth muscle of the ductus deferens, the accessory male glands, and the skeletal muscle around the bulb of the penis. These rhythmic contractions expel sperm cells and the products of the male accessory glands (prostate, seminal vesicles, and bulbourethral glands) along the penis and out of the external urethral orifice. The mature penis has a mobile foreskin that covers the glans of the penis and facilitates penile sliding during intercourse. The foreskin may be removed for reasons of social custom or religious tradition in a procedure called circumcision.

▶ **STRUCTURE AND FUNCTION OF THE PENIS**

The penis consists of three bodies of erectile tissue that can fill with blood under pressure. The paired corpora cavernosa of the body of the penis (singular—corpus cavernosum) are continuous with the crura of the base of the penis (singular—crus) and develop the highest pressure and greatest rigidity during erection. The single corpus spongiosum of the body of the penis is continuous with the bulb of the base of the penis and carries the spongy or penile urethra. The corpus spongiosum and bulb do not become as rigid during erection because the ejaculate must pass down the urethra in their center to the penile tip.

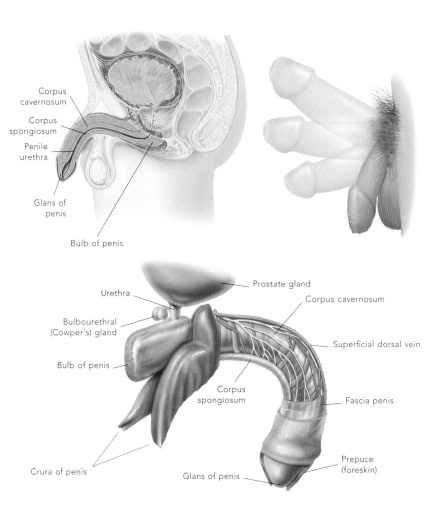

Corpus cavernosum

Corpus spongiosum

Penile urethra

Glans of penis

Bulb of penis

Urethra

Bulbourethral (Cowper's) gland

Bulb of penis

Crura of penis

Prostate gland

Corpus cavernosum

Superficial dorsal vein

Fascia penis

Corpus spongiosum

Glans of penis

Prepuce (foreskin)

Sperm meets oocyte: fertilization

FERTILIZATION IS THE FUSION OF A SPERM CELL WITH A SECONDARY OOCYTE TO FORM A ZYGOTE. Sperm cells are deposited at the uterine cervix during ejaculation, but few reach the interior of the uterus. Most sperm are destroyed by the acidic environment of the vagina, and others fail to penetrate the mucus plug in the cervical canal. Still other sperm cells can be destroyed by the woman's immune system.

The stage of a woman's cycle is also a critical determinant of the successful movement of sperm: when the woman is in the proliferative phase, estrogen is high so the cervical mucus is thin and watery and it is much easier for sperm to move through the female reproductive tract; but when the woman is in the secretory phase, progesterone is high and the mucus is thick and difficult to penetrate.

Oocytes are often in the ampulla of the uterine tube when they are fertilized, so sperm cells must climb that far. Due to the time it takes for sperm to reach the uterine tube, fertilization is usually successful when sperm are deposited into the vagina in a three-day window: two days before ovulation to one day after.

When a sperm cell encounters the oocyte, the enzymes in the acrosomal head of the sperm cell break down the protein barrier around the oocyte. The sperm cell binds to the plasma membrane of the oocyte and releases the sperm nucleus into the oocyte. The sperm nucleus swells to become a male pronucleus and combines with the female pronucleus of the oocyte. The chromosomes of the two pronuclei are mixed to form the nucleus of the new embryo.

▶ **THE PROCESS OF FERTILIZATION**
The process of fertilization is the fusion of a sperm cell and the egg cell to form a zygote (pre-embryo). When the head of the sperm comes into contact with the corona radiata, the enzymes in the acrosome are released to break through the corona radiata and zona pellucida. The successful sperm penetrates the zona pellucida, and its nucleus is released into the interior of the egg to form a male pronucleus that combines with the female pronucleus to form the nucleus of the zygote.

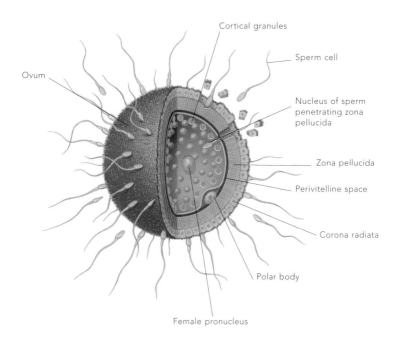

Cortical granules

Sperm cell

Ovum

Nucleus of sperm
penetrating zona
pellucida

Zona pellucida

Perivitelline space

Corona radiata

Polar body

Female pronucleus

Fertility & contraception

FERTILITY, OR THE ABILITY TO CONCEIVE, depends on many diverse factors working optimally. Infertility is the inability to produce a pregnancy after one year of unprotected sexual intercourse. Common causes of infertility include insufficient or poorly motile sperm, the presence of anti-sperm antibodies in the female reproductive tract, and the obstruction of the female reproductive tract due to previous pelvic inflammatory disease.

There are several groups of birth control methods: behavioral, barrier, hormonal, intrauterine, and permanent (sterilization). Behavioral methods depend on choosing not to have sexual intercourse when the woman is at her most fertile. In other words, this is a method of periodic abstinence based on knowing when ovulation occurs, e.g., the rhythm method. Barrier methods depend on placing some form of physical barrier between the sperm and oocyte, e.g., a condom or cervical cap. Barrier methods are usually used in conjunction with a spermicide.

Hormonal methods like oral contraceptives depend on the use of

▶ **METHODS OF CONTRACEPTION**
Intrauterine methods of contraception include the intrauterine device (IUD). The presence of a foreign body in the form of a coiled loop, "T" shape, or "7" shape, particularly with the presence of copper, prevents the implantation of the early embryo in the uterine wall.

synthetic estrogen and progesterone to induce negative feedback on the hypothalamus and the anterior pituitary to circumvent the release of the hormones required for ovulation. Intrauterine methods depend on the introduction of foreign material (an intrauterine device) into the uterus to prevent implantation of the early embryo. Copper may be added to the device to increase effectiveness. Permanent methods rely on a surgical intervention to prevent sperm cells from being ejaculated (vasectomy) or to prevent sperm from reaching the oocyte (tubal ligation).

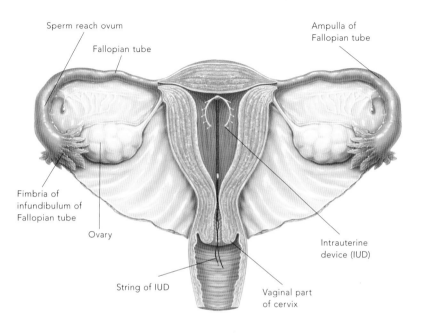

Sperm reach ovum

Fallopian tube

Ampulla of
Fallopian tube

Fimbria of
infundibulum of
Fallopian tube

Ovary

Intrauterine
device (IUD)

String of IUD

Vaginal part
of cervix

Function of the placenta

THE PLACENTA IS THE ORGAN THAT SUPPORTS THE DEVELOPING HUMAN BEYOND ABOUT WEEK 8 OF INTRAUTERINE DEVELOPMENT. The placenta is the site of exchange of oxygen, nutrients, and waste products between the mother and fetus. It is also an important organ for the production of hormones that support the pregnancy beyond week 12. The umbilical cord connects the center of the placenta to the fetus and carries a single umbilical vein, carrying oxygenated blood from the placenta to the fetus, and two umbilical arteries, carrying deoxygenated blood from the fetus to the placenta. Some substances cross the placental barrier by simple diffusion down a concentration gradient, e.g., oxygen and carbon dioxide. Other substances must be transported by specialized carriers, e.g., nutrients like glucose and amino acids.

The placenta also functions as an endocrine organ. As the pregnancy enters the third month, the placenta takes over the role of producing estrogen and progesterone from the corpus luteum of the ovary. The placenta also produces human placental lactogen and placental prolactin, which prepare the mammary gland for lactation after birth. The placenta manufactures the hormone relaxin that loosens the joints and ligaments of the pelvis so that the fetus can traverse the birth canal more easily.

If the placenta grows in the wrong part of the uterus, e.g., over the opening of the cervix, the result can be catastrophic because the process of labor will cause the separation of the placenta from the uterine wall before the fetus can be safely born.

▶ **STRUCTURE AND FUNCTION OF THE PLACENTA**

The placenta is an organ formed by the embryo to obtain nutrients from the wall of the maternal uterus. Key elements are the chorionic villi, which contain fine branches of the fetal vessels (branches of the umbilical arteries) and which are bathed in oxygen and nutrient-rich maternal blood. Gas and nutrient exchange occur between the fetal circulation and maternal blood, and the fetal blood returns along the umbilical vein.

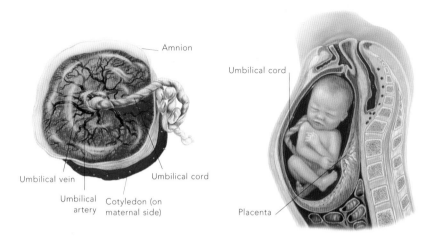

Amnion

Umbilical cord

Umbilical vein

Umbilical artery

Cotyledon (on maternal side)

Umbilical cord

Placenta

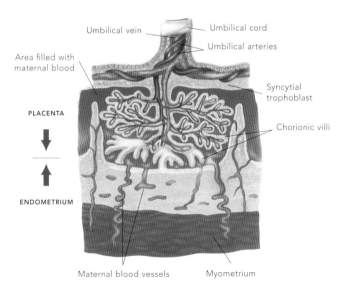

Umbilical vein

Umbilical cord

Umbilical arteries

Area filled with maternal blood

Syncytial trophoblast

PLACEMENT PLACENTA

Chorionic villi

ENDOMETRIUM

Maternal blood vessels

Myometrium

Regulation of pregnancy

THE PREGNANCY MUST BE MAINTAINED THROUGHOUT 40 WEEKS BY A CHANGING HORMONAL MILIEU. In the first two weeks of development, the corpus luteum of the ovary produces the progesterone that promotes the secretion of nutrients (uterine milk) from the endometrial glands. The developing embryo must implant in the uterine wall and start producing human chorionic gonadotrophin (hCG), which maintains the corpus luteum. If the embryo doesn't implant, the corpus luteum will degenerate and the next menstrual cycle begins.

Secretion of hCG continues for two months and inhibits the initiation of any new menstrual cycles. After two months, the production of hCG declines because the corpus luteum is no longer required. After this point, the placenta takes over as the source of estrogen and progesterone, maintaining the uterine endometrium until the time of birth.

Changes in the maternal body during pregnancy include: expansion of the uterus to accommodate the growing fetus and increase in the thickness of the uterine wall; growth of glandular tissue in the breasts and the development of additional glandular alveoli under the influence of prolactin; increase in cardiac output and blood volume but slight lowering of the red blood cell population (lower hematocrit); increase in the sensitivity of the medullary carbon dioxide sensor so that rate and depth of lung ventilation is increased; increased dietary requirements for protein, calories, calcium, iron, and folic acid; 50% increase in glomerular filtration rate in the kidneys; and pigmentation around the eyes and over the cheekbones (chloasma).

There are also significant anatomical adjustments associated with pregnancy. These include the loosening of pelvic structures under the effects of relaxin and the shifting of the abdominal and pelvic organs to accommodate the expanding uterus. The shifting of the rib cage can limit breathing.

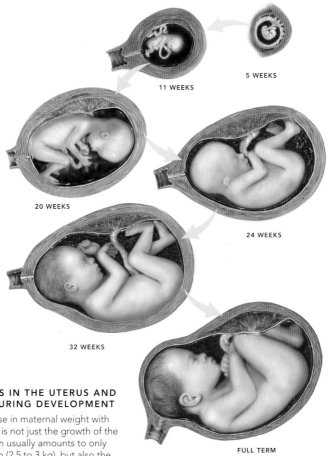

5 WEEKS

11 WEEKS

20 WEEKS

24 WEEKS

32 WEEKS

FULL TERM

▶ **CHANGES IN THE UTERUS AND FETUS DURING DEVELOPMENT**

The increase in maternal weight with pregnancy is not just the growth of the fetus, which usually amounts to only 5.5 to 6.6 lb (2.5 to 3 kg), but also the increased mass of the uterine wall (more than 2.2 lb/1 kg) and the placenta, as shown here. By the end of pregnancy, the uterus and its contents usually weigh 22 lb (10 kg).

Mechanisms & control of parturition

PARTURITION IS THE PROCESS OF GIVING BIRTH, and its initiation is actually under the control of the fetus. As the fetal adrenal gland matures, it produces cortisol, which stimulates the maternal placenta to secrete high levels of estrogen. This high level of estrogen makes the uterine smooth muscle cells (myometrium) develop receptors for oxytocin on their surfaces. Circulating oxytocin makes the myometrium twitchy and irritable, leading to irregular Braxton-Hicks contractions.

As term approaches, both the fetal and maternal hypothalami make more oxytocin, which also stimulates the placenta to make prostaglandins. The prostaglandins dilate the uterine cervix and work with the circulating oxytocin to increase the strength and frequency of uterine contractions. There is a positive feedback such that stretching of the uterine cervix by the uterine contractions stimulates the release of even more oxytocin and prostaglandins so that uterine contractions continue to grow.

Labor itself has three stages. The first stage is the dilation stage, when the cervix gradually dilates to 4 inches (10 cm). This is achieved by waves of contractions spreading from the upper parts of the uterus toward the cervix, gradually pulling the cervical tissue upward and thinning it (effacement). When the cervix is fully dilated, the expulsion stage begins. Strong uterine contractions push the fetal head into the vaginal birth canal and out to the external environment. The emergence of the fetal head and body along with a length of umbilical cord marks the end of the expulsion stage. The final stage is the placental stage, when the placenta (afterbirth) peels away from the uterine wall and is expelled to the exterior.

▶ **THE PROCESS OF PARTURITION**
Parturition begins with dilation of the cervix (first stage), which can require many hours. Once the cervix is dilated (middle left image), the head and body can move through the birth canal and be delivered (second stage). Expulsion of the placenta is the third stage.

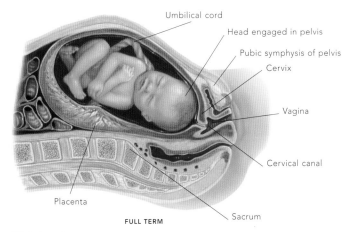

Umbilical cord

Head engaged in pelvis

Pubic symphysis of pelvis

Cervix

Vagina

Cervical canal

Placenta

Sacrum

FULL TERM

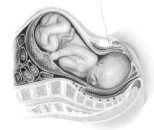

Dilated cervix

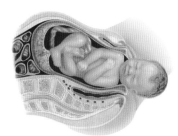

PRESENTATION OF HEAD

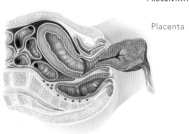

Placenta

EXPULSION OF PLACENTA

Mammary gland growth & lactation

THE MAMMARY GLANDS ARE PART OF THE INTEGUMENTARY SYSTEM, but their function is intimately involved with reproduction. Mammary glands produce milk for the newborn and infant, but they are actually modified sweat glands. Each mammary gland consists of 15 to 25 lobes that radiate from a central nipple and areola. The lobes are divided into lobules, which in turn contain glandular alveoli that produce milk when a woman is lactating. Myoepithelial cells surround the alveoli and contract in response to circulating oxytocin to expel milk. Milk passes down the lactiferous ducts and accumulates in the lactiferous sinuses immediately behind the areola before the milk is expelled from the nipple during suckling.

In the last few months of pregnancy, placental estrogen, progesterone, and human placental lactogen stimulate the production of prolactin from the anterior pituitary. Prolactin and estrogen stimulate the growth of alveoli in the breast and the branching of lactiferous ducts, preparing the breast for lactation.

The breasts secrete a thick yellowish fluid (colostrum) that is rich in protein and secretory antibodies during the first few days after delivery. This is to provide passive immunity for the newborn and protect against pathogens that cause diarrhea. True breast milk starts a few days after birth and is expelled from the breast by a let-down reflex. This is under control of oxytocin from the hypothalamus and posterior pituitary. Suckling of the nipple stimulates sensory nerves to the spinal cord, which conveys this information to the hypothalamus. The sensory stimulation triggers oxytocin release to cause the contraction of the myoepithelial cells and eject milk from the breast.

▶ **STRUCTURE OF THE BREAST**

Milk is produced in the lobules of the mammary gland. It flows along the lactiferous ducts to collect in the lactiferous sinuses deep to the areola. Suckling of the nipple by the infant stimulates ejection of milk from the lactiferous sinuses and out of the nipple. The structure of the breast is supported by fibrocartilaginous septa (suspensory ligaments).

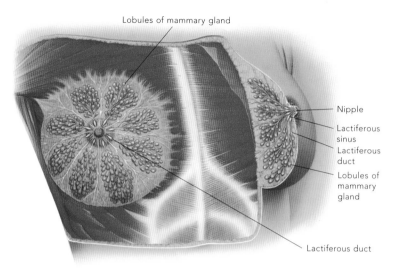

Lobules of mammary gland

Nipple

Lactiferous sinus

Lactiferous duct

Lobules of mammary gland

Lactiferous duct

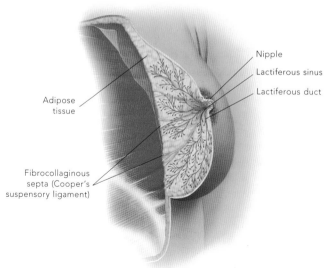

Nipple

Lactiferous sinus

Lactiferous duct

Adipose tissue

Fibrocollaginous septa (Cooper's suspensory ligament)

Sex determination & development during prenatal life

DEVELOPMENT OF THE SEXUAL ORGANS DEPENDS INITIALLY ON THE CHROMOSOMES PRESENT IN THE EMBRYO'S CELLS, in both sexes. The normal human karyotype (the collection of chromosomes) is 22 pairs of nonsex chromosomes (autosomes) and two sex chromosomes (karyotype of 46,XX for females and 46,XY for males).

The precursors of sex cells are the primordial germinal cells that first appear in the yolk sac wall of the embryo at four weeks. By six weeks of embryonic life, these germinal cells migrate to the posterior body wall to form cell clumps in the genital ridges, from which the testes and ovaries will develop. Initially, these indifferent gonads have an outer cortex and an inner medulla in both sexes, but after seven weeks, the testes and ovaries begin to follow separate developmental paths.

In males the cortex degenerates and the medulla develops into the testis. This process is under the control of a testis-determining factor (TDF) that is coded for on the sex-determining region of the Y chromosome (SRY). By eight weeks' development, the Leydig cells of the fetal testis begin to produce testosterone to drive the development of external genitalia (see illustrations). The testes will migrate under guidance from the gubernaculum to the developing scrotal swelling, arriving there by the time of birth.

In females, the medulla of the indifferent gonad degenerates and the cortex develops into the ovary. The primordial germinal cells divide to produce the primary oocytes, so the number of oocytes a woman will produce during her reproductive life is set in early fetal life. In the absence of androgens, the embryonic genital tubercle becomes a clitoris and the labioscrotal swellings form into labia, flanking a vaginal opening.

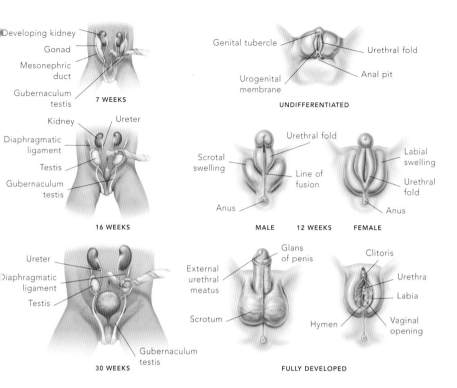

7 WEEKS

- Developing kidney
- Gonad
- Mesonephric duct
- Gubernaculum testis

UNDIFFERENTIATED

- Genital tubercle
- Urogenital membrane
- Urethral fold
- Anal pit

16 WEEKS

- Kidney
- Ureter
- Diaphragmatic ligament
- Testis
- Gubernaculum testis

MALE   12 WEEKS   FEMALE

- Urethral fold
- Scrotal swelling
- Line of fusion
- Anus
- Labial swelling
- Urethral fold
- Anus

30 WEEKS

- Ureter
- Diaphragmatic ligament
- Testis
- Gubernaculum testis

FULLY DEVELOPED

- Glans of penis
- External urethral meatus
- Scrotum
- Clitoris
- Urethra
- Labia
- Hymen
- Vaginal opening

FULLY DEVELOPED

- Ureter
- Bladder
- Vas deferens
- Epididymis
- Testis

▲ DEVELOPMENT OF THE GENITALIA

In males the testes develop on the posterior abdominal wall and migrate to the scrotum during fetal life (left-hand column of illustrations). The external genitalia of the two sexes (right-hand illustrations) develop by differential growth of the genital tubercle (to form either a penis or clitoris) and the genital swellings (to form either a scrotum or labia majora).

Intersex & sex chromosomal variations

INTERSEX IS DEFINED AS VARIATIONS IN SEX CHARACTERISTICS THAT DO NOT FIT THE USUAL BINARY DIVISION INTO MALE OR FEMALE. There are variations in the development of sexual organs that arise due to either chromosomal differences and/or genetically determined alterations in tissue sensitivity to hormones.

Klinefelter's syndrome is seen in males with an extra X chromosome, i.e., with the karyotype 47,XXY. Someone with this syndrome will be phenotypically male, i.e., will have male external genitalia due to the presence of the Y chromosome, but will have small testes, low testosterone levels, and high estrogen levels. The levels of FSH in the blood will be high because testosterone is low and there is no negative feedback on the hypothalamus and anterior pituitary. The high estrogen will cause breast development (gynecomastia).

In Turner's syndrome (karyotype 45,XO), the absence of a second X chromosome means that the ovaries fail to develop. There will be little or no estrogen production, so there is no negative feedback on the hypothalamus and anterior pituitary, and levels of FSH in the blood will be high. People with Turner's syndrome are phenotypically female but do not show any female secondary sexual characteristics, i.e., no breast or pubic hair development. They are short with a broad chest and webbing of the neck.

Other syndromes may involve disorders of the chemical signaling that produces normal genitalia during embryonic life. In androgen insensitivity syndrome (AIS), androgens do not exert their normal effect during embryonic life and a chromosomally normal male fails to develop external male genitalia, while the testes remain inside the abdomen. In Müllerian agenesis (Rokitansky-Küster-Hauser syndrome), the uterus, cervix, and upper vagina are absent.

▶ **SEX CHROMOSOME VARIATION**
Turner's syndrome is caused by the absence of a second X chromosome so there is only a single X chromosome. This has the effect of people being phenotypically female but without any secondary sexual characteristics.

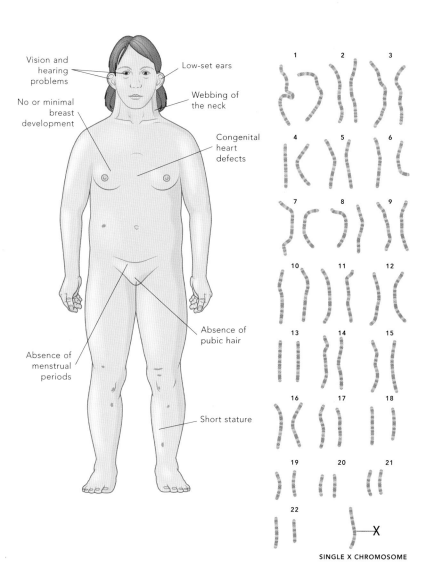

Vision and hearing problems

Low-set ears

No or minimal breast development

Webbing of the neck

Congenital heart defects

Absence of pubic hair

Absence of menstrual periods

Short stature

1 2 3
4 5 6
7 8 9
10 11 12
13 14 15
16 17 18
19 20 21
22

X

SINGLE X CHROMOSOME

Puberty & sexual maturation

PUBERTY BEGINS FOR MOST GIRLS AT AROUND 9 TO 11 YEARS OF AGE, although this varies with the level of nutrition. The key event that triggers puberty is the production of more estrogen and progesterone from the ovaries. Just before puberty, the hypothalamus becomes less sensitive to negative feedback by estrogen and progesterone, so levels of gonadotrophin-releasing hormone rise, in turn increasing follicle-stimulating hormone and luteinizing hormone production from the anterior pituitary and driving more production of estrogen and progesterone from the ovaries.

These hormones produce the secondary sexual characteristics of mature females, so the breasts begin to grow, pubic and armpit hair appears, and fat tissue is distributed in the feminine pattern (in the breasts and around the hips and thighs). The skeleton also grows, increasing height and widening the pelvis, and increased sebaceous gland activity in the skin can result in acne. The first episode of menstrual bleeding (menarche) usually occurs about two years after the onset of puberty.

In males, secretion of testosterone from the testes begins to rise when the sensitivity of the hypothalamus to negative feedback declines at around 11 years of age. Puberty begins between the ages of 12 and 14, when males develop secondary sexual characteristics, such as growth of pubic, armpit, and facial hair; enhanced body hair generally; increased muscle bulk and skeletal growth; enlargement of the larynx and deepening of the voice; enlargement of the penis; thickening of the skin; and increased sebaceous gland activity (causing acne). Spermatogenesis (sperm production) begins a few years after the onset of puberty.

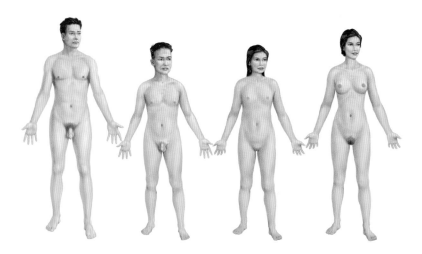

17-YEAR-OLD MALE
Appearance of facial hair along with body and pubic hair, and increase in the size of the testes and penis are the predominant signs of male puberty. The larynx also increases in size, creating the "Adam's apple" at the front of the neck.

12-YEAR-OLD MALE
Puberty in males occurs about two years later than in females. At age 12, a boy has little or no body hair, has a small penis and testes, and still appears childlike.

10-YEAR-OLD FEMALE
There is little or no breast development, no pubic hair, and an almost boyish appearance at this age.

17-YEAR-OLD FEMALE
Fully developed breasts, hair growth around the vulva, and widening of the hips are the main outward signs of female puberty.

▲ PHYSICAL CHANGES AT PUBERTY

Puberty in both sexes is manifested externally by the appearance of secondary sexual characteristics, e.g., increased muscle mass, penile growth, deeper voice, and pubic hair in males and growth of breasts, deposition of fat in feminine distribution, and growth of pubic hair in females.

Metabolism of carbohydrates

CARBOHYDRATES IN FOOD MAY BE IN THE FORM OF MONOSACCHARIDES AND DISACCHARIDES, i.e., simple sugars in fruit, dairy products, sugar cane, and candy, or polysaccharides, i.e., complex carbohydrates in rice, pasta, bread, potatoes, vegetables, and glycogen in meat. Complex carbohydrates like maltose and glycogen are broken down to simple sugars, which are absorbed across the wall of the small intestine.

Some carbohydrates known as fiber are indigestible because humans lack the necessary enzymes to break down cellulose. Insoluble fiber, such as is found in whole grains, fruit skin, bran, and some vegetables, cannot be digested at all and passes through the gut unaltered. Soluble fiber, such as is found in fruit pulp, oats, and vegetables, is partially digestible by gut bacteria and can have beneficial effects such as lowering cholesterol in the blood.

Current recommendations are that 45% to 65% of nutrients should come from carbohydrates and that these should mostly be polysaccharides. Glucose is an essential nutrient for the brain, and the maintenance of a stable blood glucose concentration is a key homeostatic goal.

Carbohydrates are not only essential energy sources—they are also used in the production of many structural molecules such as glycoproteins. On the other hand, a diet with an excess of highly processed food loaded with simple sugars is nutritionally poor ("empty calories") and leads to obesity. The concentration of glucose in the blood is regulated by the hormones insulin and glucagon, which act in tandem to keep blood sugar in an optimal range of 72 to 126 mg/dL (4 to 7 mM).

▸ **CARBOHYDRATE METABOLISM IN THE BODY**

Carbohydrates are a major energy source for the body and are particularly important for brain metabolism. Carbohydrates are absorbed in the small intestine and stored in the liver and skeletal muscles as glycogen. The pancreas is important as a source of the enzymes that allow carbohydrate digestion (amylases and disaccharidases from the exocrine pancreas) and as the site of the pancreatic islet cells that make the hormones insulin and glucose for the regulation of blood sugar level.

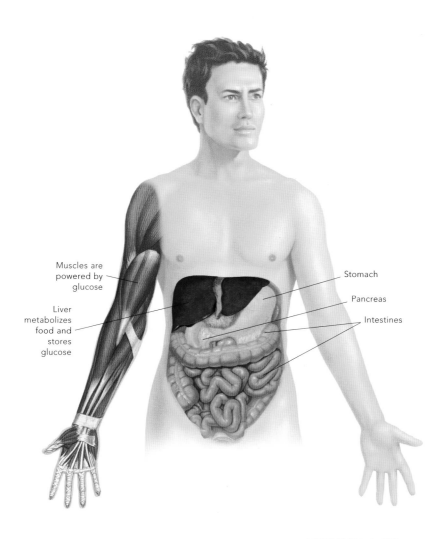

Muscles are powered by glucose

Liver metabolizes food and stores glucose

Stomach

Pancreas

Intestines

Hormonal regulation of blood glucose levels

BLOOD GLUCOSE LEVEL IN A FASTING INDIVIDUAL SHOULD BE LESS THAN 100 mg/dL (about 5.5 mM). A fasting blood glucose level above 125 mg/dL may indicate diabetes mellitus. Blood glucose levels are maintained in balance by an interplay between the effects of two hormones: insulin and glucagon.

Rising blood glucose levels, such as after eating a carbohydrate-rich meal, stimulate the secretion of insulin from beta cells of the pancreatic islets. Insulin promotes the uptake of glucose by the body's cells and stimulates the liver to convert glucose to glycogen stores. In a normal individual, these two actions will lower blood glucose levels back to normal. If the person has insulin resistance, the blood glucose levels stay high.

If blood glucose levels fall, e.g., due to missing a meal, glucagon is released from the alpha cells of the pancreatic islets. Glucagon stimulates the liver to convert glycogen stores to glucose and engage in gluconeogenesis (manufacture of new glucose from amino acids). This returns the blood glucose level to normal. Release of epinephrine and norepinephrine from the adrenal medulla in short-term emergency situations may also increase blood glucose levels by stimulating glucose release from the liver.

▶ **REGULATION OF BLOOD GLUCOSE LEVELS**

The optimal concentration of glucose in the blood is maintained by two hormones with largely opposite effects. The first, insulin is released by the beta cells of the pancreatic islets in response to elevated blood glucose and stimulates the incorporation of glucose into cells in general and carbohydrate stores in particular. The second, glucagon is released from alpha cells of the pancreatic islets in response to low blood glucose concentration and mobilizes glucose from carbohydrate stores and other sources.

CONTROL OF BLOOD GLUCOSE LEVEL

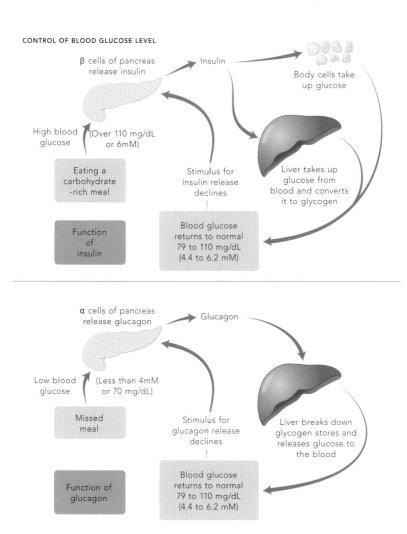

β cells of pancreas release insulin → Insulin → Body cells take up glucose

High blood glucose (Over 110 mg/dL or 6mM)

Eating a carbohydrate -rich meal

Function of insulin

Stimulus for insulin release declines

Liver takes up glucose from blood and converts it to glycogen

Blood glucose returns to normal 79 to 110 mg/dL (4.4 to 6.2 mM)

α cells of pancreas release glucagon → Glucagon

Low blood glucose (Less than 4mM or 70 mg/dL)

Missed meal

Function of glucagon

Stimulus for glucagon release declines

Liver breaks down glycogen stores and releases glucose to the blood

Blood glucose returns to normal 79 to 110 mg/dL (4.4 to 6.2 mM)

Metabolism of fats, cholesterol, & lipoproteins

THERE ARE MANY TYPES OF FATS (lipids) ingested in the diet. Key types include triglycerides, cholesterol, and fat-soluble vitamins (A, K, D). Saturated fatty acids are hydrocarbon chains with no double bonds between carbon atoms. They are mainly found in animal fats and should not make up more than 10% of the total energy intake. Unsaturated fats have hydrocarbon chains with one or more double bonds between the carbon atoms.

Fats are important nutrients and can be used as energy sources with more than twice the energy content per gram of carbohydrates and proteins. They are also an important part of structural lipids in the body. Essential fatty acids, e.g., linoleic and linolenic acids, cannot be made by the body and must be ingested to build the body's structural lipids, and omega-3 fatty acids have demonstrated positive effects on cardiovascular health.

Dietary fats must be broken down into fatty acids and glycerol to be absorbed across the gut wall. This requires enzymes called lipases that are found in the pancreatic juice and in the lining of the small intestine. Constituents of fats like cholesterol are hydrophobic (water-insoluble) and are transported in the blood packaged along with other fats into carrier proteins called lipoproteins.

There are several different types of lipoproteins: very low-density lipoproteins (VLDL), low-density lipoproteins (LDL), and high-density lipoproteins (HDL). VLDL are triglyceride-rich and are converted to fatty acids and glycerol by the enzyme lipoprotein lipase in endothelial cells. LDL are the main lipoproteins for transporting cholesterol, so elevated LDL in the blood can be seen as an indication for the prescription of cholesterol-lowering medication. HDL transfer lipids from the peripheral tissues of the body toward the liver. VLDL and LDL are referred to as "bad" cholesterol because high levels are associated with accelerated arterial disease. On the other hand, HDL is "good" cholesterol because it reduces fat deposition in arteries.

▶ STRUCTURE OF A LIPOPROTEIN

Lipoproteins are specialized structures for transporting water-insoluble fats around in the watery environment of the blood. They have a core of neutral hydrophobic lipids surrounded by a shell of charged phospholipids. The shell also contains embedded proteins.

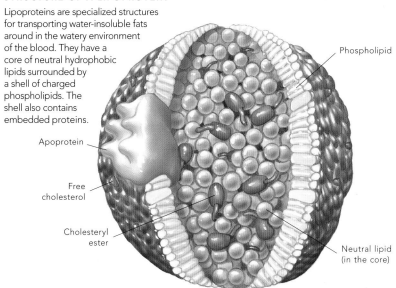

Phospholipid

Apoprotein

Free cholesterol

Cholesteryl ester

Neutral lipid (in the core)

ATHEROSCLEROSIS

Atherosclerosis is a common condition in western society in which fatty/fibrous plaques develop in the arterial wall. These can cause arterial blockage or rupture, and the condition is particularly serious when it occurs in arteries supplying the brain and heart. Blockage of cerebral arteries can cause stroke. Blockage of heart (coronary) arteries causes chest pain on exertion (angina pectoris) and may cause death of heart muscle (myocardial infarction). Risk factors include genes, male gender, fatty diet, high blood pressure, high blood lipids, smoking, excess alcohol intake, inadequate exercise, and obesity.

Metabolism of amino acids & proteins

PROTEINS AND AMINO ACIDS ARE IMPORTANT AS SOURCES OF ENERGY, as building blocks for structural proteins, and for body metabolism in the form of enzymes. Good dietary sources of protein are meat; dairy products; eggs; and plant proteins from legumes, nuts, seeds, some grains, and soy.

Twenty amino acids make up the body's proteins, and of these 11 are said to be nonessential because the liver can make them from other molecules. The nine essential amino acids—histidine, isoleucine, leucine, lysine, methionine, phenylalanine, threonine, tryptophan, and valine—must be provided by the diet.

Current recommendations are that protein makes up between 10% and 35% of energy intake. Dietary proteins and amino acids contribute to an amino acid pool, which can be used to build body proteins, e.g., contractile and structural proteins of muscles, or structural proteins of the brain; be converted to amino acid-based body chemicals, e.g., hormones like thyroxine, nucleotides for DNA, and creatine; or provide an energy source.

Amino acids can be converted to glucose or fat for energy production or storage. Before amino acids can be used for energy production, the nitrogen must be removed in a biochemical reaction called transamination. This process occurs in the liver and produces carbon skeletons that can be used to produce adenosine triphosphate (ATP) and the waste product ammonia (NH_3). The ammonia is highly toxic and is converted to urea for excretion in the urine.

▶ USE OF THE AMINO ACID POOL
The daily intake of pure dietary protein (between 1.5 and 2.5 oz/45 and 70 g) contributes to an amino acid pool that can be used for a variety of purposes, ranging from structural proteins like collagen, to nitrogen-containing amino acid derivatives like hormones and nucleotides, to carbon skeletons for energy production.

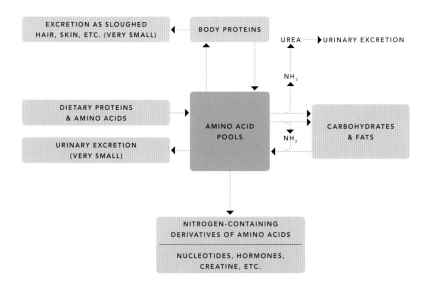

Micronutrients

APART FROM CARBOHYDRATES, PROTEIN, AND FATS, A RANGE OF MICRONUTRIENTS—vitamins and minerals like sodium, potassium, iron, calcium, magnesium, iodine, and selenium—are needed for health. These may be essential micronutrients in the sense that they cannot be made by the body or can only be made in small quantities by the body or the gut flora.

Many vitamins are essential for metabolic processes, and their deficiency can have serious health consequences, e.g., pellagra from B3 or niacin deficiency, and beri-beri from B1 or thiamine deficiency. Some vitamins are lipid-soluble (A, D, E, K) and require the ability to absorb fats, but most are water-soluble (B complex, C).

Vitamin A (retinal or retinoic acid) is important for formation of visual pigments and the growth and reproduction of epithelial cells. Members of the B complex of vitamins, i.e., thiamine or B1, riboflavin or B2, niacin or B3, pantothenic acid or B5, pyridoxine or B6, cyanocobalamin or B12, folic acid, and biotin, play key roles in carbohydrate metabolism, protein synthesis, fat, and DNA synthesis. Vitamin C is essential for collagen manufacture and is stored for only three weeks in the body. A deficiency of vitamin C due to insufficient fresh fruit and vegetables in the diet leads to scurvy. Vitamin D is vital for absorption of calcium from the gut. It can be obtained from dairy products and oily fish or made through exposure of the skin to sunlight. Vitamin E is an important antioxidant and may prevent cellular damage. Vitamin K (phylloquinone) is essential for the production of key clotting factors (III, VII, IX, X), and a deficiency can lead to fatal hemorrhage.

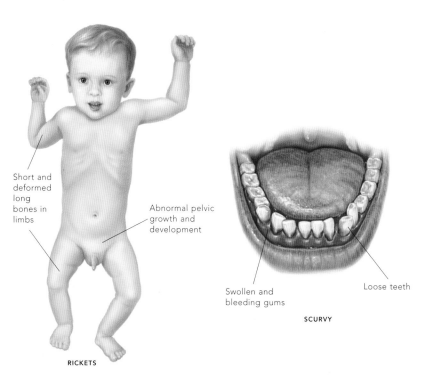

Short and deformed long bones in limbs

Abnormal pelvic growth and development

Swollen and bleeding gums

Loose teeth

SCURVY

RICKETS

▲ EFFECTS OF MICRONUTRIENT DEFICIENCY

The effects of vitamin deficiency can be serious. Vitamin D is essential for absorption of calcium from the gut. Deficiency of vitamin D (whether due to diet or lack of sun-exposure) classically leads to rickets (left illustration), a condition in which there is defective mineralization of bones resulting in deformities of weight-bearing long bones and the pelvis. Vitamin C is essential for production of hydroxyproline, a key amino acid in the structural protein collagen. Defective collagen formation causes weak brittle bones due to impaired formation of the organic part of bone, damage to epithelium due to weakened connective tissues, and loss of the teeth due to defective formation of the periodontal ligament fibers.

Thermoregulation

THERMOREGULATION IS THE PROCESS OF MAINTAINING A CONSTANT INTERNAL OR CORE BODY TEMPERATURE AT AROUND 99.5°F (37.5°C). Thermoregulation depends on many factors, both internal and external, because the balance between metabolic rate and heat loss to the external environment determines the core temperature. Thermoregulation requires thermoreceptors to detect the body's temperature, and these are located in both the skin and the hypothalamus. The hypothalamic regulatory centers compare the current temperature with the thermoregulatory set point.

If core temperature is above the set point, then the hypothalamic heat-loss center will command the sympathetic nervous system to increase blood flow to the skin to dump core heat to the external environment by radiation. The sympathetic nervous system will also stimulate sweat glands to increase sweat production so heat is lost to the external environment by evaporation.

If body temperature falls, thermoreceptors in the skin and hypothalamus detect the change and the heat-promoting center in the hypothalamus is activated. This instigates contraction of the blood vessels in the skin to minimize loss of core heat at the skin surface and causes elevation of the body hairs by the arrector pili muscles in the skin to keep a layer of insulating air close to the skin surface. Shivering in the skeletal muscles is activated to produce more heat, and the general metabolic rate of the body is increased. Metabolic rate is increased by the release of thyroid hormones from the thyroid gland and the discharge of epinephrine and norepinephrine from the adrenal medulla.

▶ **MECHANISMS FOR REGULATING BODY TEMPERATURE**

Since the main route for heat loss from the body is via the skin, most of the mechanisms for thermoregulation are located there. These include the hairs of the skin that can be raised for insulation, sweat glands that promote heat loss by evaporation, and dermal vasculature that can be dilated or constricted as required.

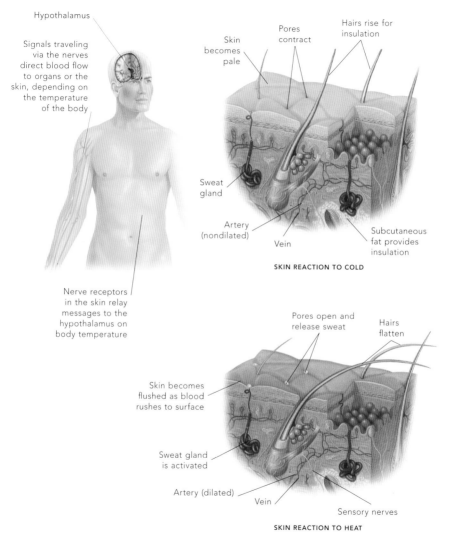

Hypothalamus

Signals traveling via the nerves direct blood flow to organs or the skin, depending on the temperature of the body

Nerve receptors in the skin relay messages to the hypothalamus on body temperature

Skin becomes pale

Pores contract

Hairs rise for insulation

Sweat gland

Artery (nondilated)

Vein

Subcutaneous fat provides insulation

SKIN REACTION TO COLD

Pores open and release sweat

Hairs flatten

Skin becomes flushed as blood rushes to surface

Sweat gland is activated

Artery (dilated)

Vein

Sensory nerves

SKIN REACTION TO HEAT

Control of metabolic rate

METABOLIC RATE IS THE TOTAL AMOUNT OF ENERGY EXPENDED BY THE BODY TO PERFORM ALL OF ITS PROCESSES. The metabolic processes of the body are powered by adenosine triphosphate (ATP) produced mainly by the oxidative phosphorylation of fuels, i.e., carbohydrates, fats, and amino acids. However, no biochemical process is 100% efficient, and ultimately all energy expended in the body becomes heat. Nevertheless, this heat energy contributes to the essential maintenance of body temperature.

Energy expenditure can change with activity—sometimes tenfold in extreme exercise, but an important concept is the basal metabolic rate, which is the minimal rate of metabolism for a waking individual. This is measured in a constant temperature environment, when the individual has not eaten for 12 hours and is not emotionally stressed or engaged in exercise. Metabolic rate is affected by a variety of factors: thyroid and growth hormone levels, fever, nutritional state, and physical activity.

Thyroid hormone level is a key determinant because thyroxine (T4) and triiodothyronine (T3) increase the basal metabolic rate of all body cells by direct actions on the mitochondria, making oxidative phosphorylation generate more heat for a given metabolic effect. Thyroid activity is regulated by a negative feedback cycle such that circulating T3 and T4 feed back on the hypothalamus and anterior pituitary to turn down the release of thyroid-stimulating hormone.

▸ **HORMONAL REGULATION OF METABOLIC RATE**

The general metabolic rate of the body is regulated by the thyroid hormones triiodothyronine (T3) and thyroxine (T4). These are both produced by the thyroid gland and increase the basal metabolic rate of all body cells. Secretion of T3 and T4 is controlled by negative feedback of those hormones on the hypothalamus and anterior pituitary gland to reduce thyrotrophin-releasing hormone (TRH) and thyroid-stimulating hormone (TSH), respectively.

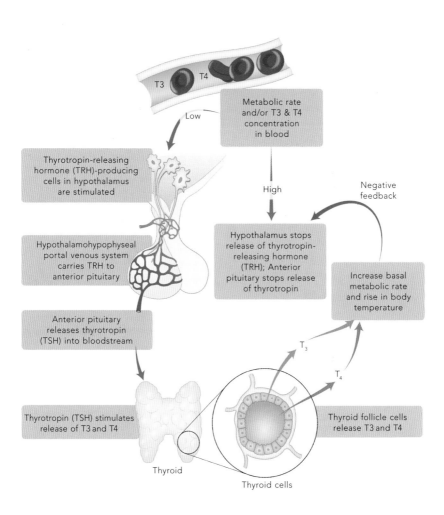

T3 T4

Metabolic rate and/or T3 & T4 concentration in blood

Low

High

Negative feedback

Thyrotropin-releasing hormone (TRH)-producing cells in hypothalamus are stimulated

Hypothalamus stops release of thyrotropin-releasing hormone (TRH); Anterior pituitary stops release of thyrotropin

Hypothalamohypophyseal portal venous system carries TRH to anterior pituitary

Increase basal metabolic rate and rise in body temperature

Anterior pituitary releases thyrotropin (TSH) into bloodstream

T_3

T_4

Thyrotropin (TSH) stimulates release of T3 and T4

Thyroid follicle cells release T3 and T4

Thyroid

Thyroid cells

Regulation of food intake

FEEDING IS CONTROLLED BY NUCLEI WITHIN THE HYPOTHALAMUS. The medial hypothalamus contains a satiety center that induces feelings of fullness and stops the desire to eat. The lateral hypothalamus contains the hunger or feeding center that stimulates feelings of hunger and the craving for more food. These centers are acted upon by a variety of hormonal and neural signals.

Fat cells produce a hormonal signal called leptin that inhibits nerve cells in the hunger center while stimulating nerve cells in the satiety center. Large quantities of body fat will therefore tend to reduce appetite. Insulin produced by the pancreas in response to nutrient loading of the body also acts on the brain to promote the feeling of satiety. Other neural signals from the stomach and small intestine, signaling fullness of those organs, are detected by stretch receptors in the gut wall and pass along the vagus nerve to the brainstem, where pathways to the hypothalamus will stimulate the satiety center and decrease appetite.

Signals for increased appetite come from the mucosal lining of the stomach. The stomach epithelium makes a hormone called ghrelin that stimulates the hunger center in the hypothalamus to produce the neurotransmitters orexin and neuropeptide, both of which promote hunger. Mechanisms by which gastric bypass surgery may reduce food intake include decreasing the volume at which the stomach sends a "full" signal and diminishing the production of ghrelin by reducing the area of stomach mucosa.

▶ **MECHANISMS IN THE REGULATION OF FOOD INTAKE**

The intake of food is regulated by the hypothalamus. Signals that influence appetite arise from the body's own fat stores (via leptin and insulin levels in the blood) and the gastrointestinal tract itself, e.g., by stretch receptor information signaling fullness of the distensible parts of the gut. Disordered appetitive behavior can lead to anorexia (reduction in body weight to a body mass index (BMI) of less than 18), which is characterized by numerous emotional and physical problems.

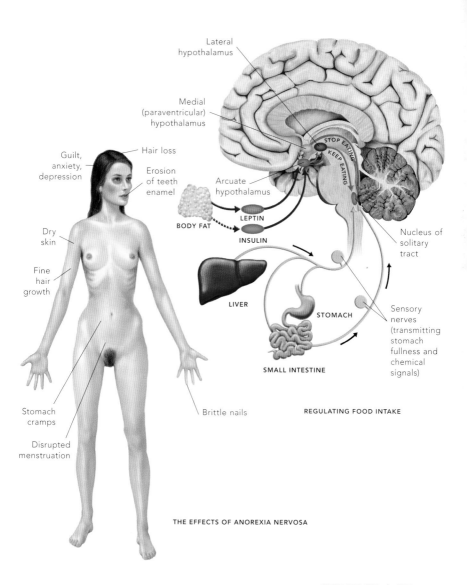

Lateral hypothalamus

Medial (paraventricular) hypothalamus

Guilt, anxiety, depression

Hair loss

Erosion of teeth enamel

Arcuate hypothalamus

STOP EATING

KEEP EATING

BODY FAT

LEPTIN

INSULIN

Nucleus of solitary tract

Dry skin

Fine hair growth

LIVER

STOMACH

Sensory nerves (transmitting stomach fullness and chemical signals)

SMALL INTESTINE

Stomach cramps

Disrupted menstruation

Brittle nails

REGULATING FOOD INTAKE

THE EFFECTS OF ANOREXIA NERVOSA

Obesity & weight control

OBESITY IS THE CLINICAL CONDITION IN WHICH THERE IS EXCESS ADIPOSE (fat) tissue for the individual's lean body mass. Body mass is usually assessed and compared by calculating the body mass index (BMI), which is the ratio of the body weight in kilograms over the height in meters squared. BMI between 18.5 and 24.9 is normal. A BMI between 25.0 and 29.9 is overweight, and over 30.0 is considered obese. An alternative measure that more closely assesses visceral (abdominal) fat is the abdominal circumference. This should be less than 40 inches (102 cm) in men and 35 inches (88 cm) in women.

Obesity can be of two types depending on what happens to the size or number of the body's fat cells. Hypertrophic obesity is when the body has a relatively normal number of fat cells, but these are swollen by as much as four times with excess amounts of fatty acids. Hypertrophic obesity is usually seen in individuals who gain weight in adulthood. Hypercellular obesity usually has its onset due to excessive weight gain during infancy or early childhood. In hypercellular obesity, the number of fat cells is increased beyond normal.

Obesity decreases a person's lifespan by six to seven years due to increased risk of chronic disease, e.g., diabetes mellitus, cardiovascular, and renal disease. The mechanism for this increased risk is not well understood but probably arises from the increased blood pressure and blood lipid concentration seen in obesity. The causes of obesity are primarily increased energy intake while leading a sedentary lifestyle, although endocrine problems like hypothyroidism and excess cortisol secretion may also cause it. Treatment is usually by reduction of calorie intake and an exercise program. Severe cases may require gastric banding or gastric bypass surgery to reduce the effective size of the stomach.

DISTRIBUTION OF EXCESS FAT IN OBESITY

When an excessive amount of chemical energy is stored as fat, it usually is found around the internal organs of the abdominal cavity. This visceral fat is a particularly strong marker for the risk of developing chronic disease, e.g., diabetes mellitus, high blood pressure, cardiac disease, and some cancers.

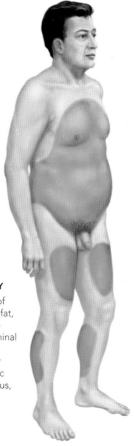

OBESITY—MALE
FAT DISTRIBUTION

OBESITY—FEMALE
FAT DISTRIBUTION

Diabetes mellitus

DIABETES MELLITUS IS A CONDITION IN WHICH THERE IS INADEQUATE CONTROL OF BLOOD GLUCOSE LEVELS. The two most common types of diabetes mellitus are known as type 1 and type 2.

Type 1 (juvenile onset, or insulin-dependent) diabetes mellitus is caused by the destruction of beta cells of the pancreas by the immune system, usually occurring during childhood or young adulthood. The resulting lack of insulin means that regardless of the level of glucose in the blood, most cells in the body are unable to bring glucose into their cytoplasm. The result is that cells essentially starve while surrounded by glucose. The result is very high levels of glucose in the blood (hyperglycemia) because glucagon is acting unopposed and making the liver engage in gluconeogenesis. Glucagon also causes increased levels of ketone bodies in the blood, and both glucose and ketones are found in the patient's urine in large quantities. The presenting symptoms are commonly thirst (polydipsia) and increased urination (polyuria) due to the osmotic effect of the elevated glucose. The only treatment at present is regular injections of insulin in combination with strict monitoring of diet and blood glucose levels.

Type 2 (or non-insulin-dependent) diabetes mellitus is a much more common condition than type 1, affecting 95% of diabetics in developed countries. It usually begins in adulthood and is strongly associated with heredity and obesity. Patients usually produce enough insulin, but the body's cells have developed insulin resistance and the beta cells do not respond adequately to increased blood glucose levels. Poorly controlled diabetes mellitus causes damage to the arteries of the body (atherosclerosis), peripheral nerves (peripheral neuropathy), the retina and lens in the eye, and the kidneys (chronic renal failure).

▶ **COMMON COMPLICATIONS OF DIABETES MELLITUS**
Diabetes mellitus has profound metabolic consequences for all tissues in the body, in particular the small vessels of the retina, kidney, and feet, as well as larger vessels of the heart and brain.

CATARACT

Cataract is a common complication of diabetes mellitus because excess glucose in the blood interferes with the metabolism of lens cells.

DIABETIC RETINOPATHY

Diabetes damages the small blood vessels in the retina of the eye. The damaged vessels can bleed, causing hemorrhages, or they can cause areas of the retina to die, resulting in loss of vision.

Blockage

DIABETIC NEPHROPATHY

Diabetic nephropathy damages the glomeruli and small blood vessels of the kidneys due to high levels of blood glucose. This results in the loss of necessary proteins through the urine, swelling of body tissues, and eventually renal failure.

ISCHEMIA

Diabetes can cause coronary artery disease, which in turn can cause death of cardiac tissue known as myocardial infarction.

FOOT ULCER

Diabetes slows the healing of body tissues and also causes degeneration of the peripheral nerves (neuropathy). These two factors act together to cause foot ulcers in diabetics.

ATHEROSCLEROSIS

Diabetes is one of the risk factors for atherosclerosis, a disease in which fatty deposits build up under the lining of the artery and block off its blood flow.

Composition & function of blood: fluid components

THE VOLUME OF CIRCULATING BLOOD IS ON AVERAGE 10 PINTS (5 liters) and makes up about 8% of the body weight. Blood consists of a liquid extracellular matrix called plasma (about 55% of the volume) and formed elements (cells and cell fragments) suspended in the plasma. About 44% of the blood volume is red blood cells (erythrocytes) that contain hemoglobin (Hb) to carry oxygen and carbon dioxide. The percentage of blood that is composed of red blood cells is called the hematocrit. The remaining 1% consists of white blood cells (leukocytes) for immune function and platelets for hemostasis (prevention of blood loss).

Blood serves many functions, which include: the transport of gases; the transport of ions, nutrients, hormones, and waste products; immune function; the maintenance of body temperature; blood clotting or hemostasis; the preservation of acid-base balance; and the stabilization of blood pressure.

The plasma contains 9% plasma proteins. Most of these are made in the liver, e.g., albumin, and dissolve in the water of the plasma to provide an osmotic pressure that tends to draw water back into the circulation at the venous end of the capillary bed. Other plasma proteins include immune system proteins, e.g., gamma globulins or antibodies, transport proteins for carrying water-insoluble molecules like fats and steroids, and clotting factor proteins for preventing blood loss following injury.

▶ **COMPONENTS OF THE BLOOD**

Blood is a connective tissue and consists of formed elements (cells and cell fragments) suspended in a fluid matrix (the plasma). Red blood cells make up most of the cellular component, but white blood cells (leukocytes) are also important for immune system function. The fluid component contains plasma proteins like albumin and clotting factors like fibrinogen.

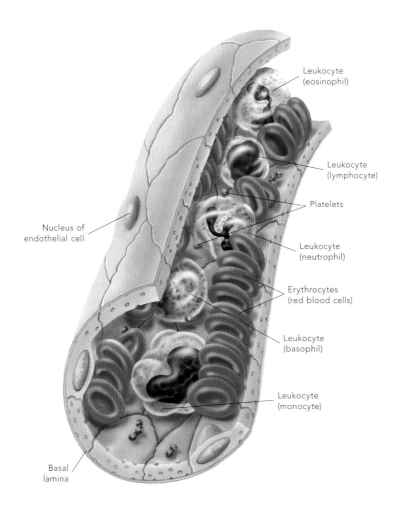

Leukocyte
(eosinophil)

Leukocyte
(lymphocyte)

Platelets

Nucleus of
endothelial cell

Leukocyte
(neutrophil)

Erythrocytes
(red blood cells)

Leukocyte
(basophil)

Leukocyte
(monocyte)

Basal
lamina

Leukemia

LEUKEMIA IS THE TERM FOR A GROUP OF DISEASES THAT ARE ESSENTIALLY MALIGNANT NEOPLASIA (literally aggressive new growth) of the white blood cells. The disease is characterized by an increase in the populations of white blood cells in the blood and in the organs that form blood, i.e., bone marrow and spleen. Leukemic cells are not able to perform the normal functions of white blood cells, i.e., fighting disease-causing organisms, so the affected individual is prone to increased incidence and severity of life-threatening infections. The invasion of the bone marrow by leukemic cells can also interfere with the formation of red blood cells and platelets, leading to anemia and problems with hemostasis.

Leukemia can be acute (developing rapidly over weeks and quickly fatal if not treated) or chronic (occurring over a period of years). Apart from overwhelming infection, acute leukemia also causes fatigue, shortness of breath, and anemia due to displacement of red blood cell production. Patients with chronic leukemia are usually older (over 50 years) and may survive for some years with the disease because the condition is usually less aggressive.

Leukemia is usually classified according to the cell line that has given rise to the abnormal cells. Acute lymphoblastic leukemia is the type of leukemia most commonly seen in children and involves cells derived from the lymphocyte cell line. Acute myeloid leukemia is the most common acute type seen in adults and involves cells derived from the granulocyte cell line. Chronic lymphocytic leukemia is the most common leukemia in people over 50 years.

▶ **EFFECTS OF LEUKEMIA**

In normal blood (top right) white blood cells make up only 0.1% of cells in the blood, but in leukemia abnormal white blood cells make up several percent of the cellular mass of the blood (top left). They may displace hematopoietic (blood-making) tissue from the red bone marrow (lower left) and infiltrate the spleen causing enlargement (bottom right).

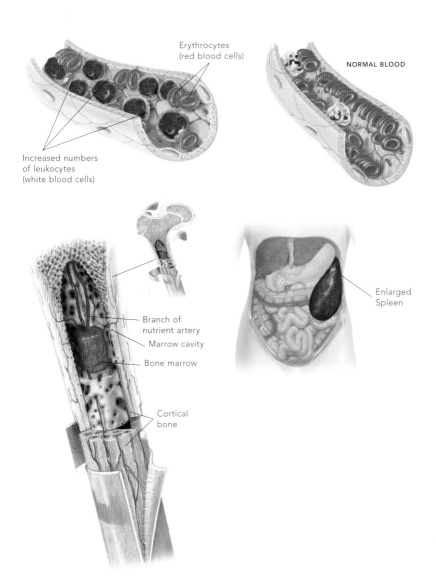

Erythrocytes
(red blood cells)

NORMAL BLOOD

Increased numbers
of leukocytes
(white blood cells)

Branch of
nutrient artery

Marrow cavity

Bone marrow

Cortical
bone

Enlarged
Spleen

Cellular components of blood

THE BULK OF THE CELLULAR COMPONENT OF BLOOD IS MADE UP OF ERYTHROCYTES OR RED BLOOD CELLS. The typical erythrocyte is a biconcave disc and lacks a nucleus and most other organelles. This biconcave or flattened doughnut shape gives erythrocytes a larger surface area for gas exchange. Each erythrocyte contains about one billion molecules of hemoglobin (Hb), which give blood its red color.

Leukocytes, or white blood cells, are immune system cells and are divided into types with granules in their cytoplasm (granulocytes like neutrophils, eosinophils, and basophils) and those without (agranulocytes like lymphocytes and monocytes). Neutrophils are the most common leukocyte (60% to 70%) and have a three- to five-lobed nucleus. They are able to migrate into tissue spaces and release their antibacterial granules or actively engulf bacteria and cellular debris. Eosinophils have a bi-lobed nucleus and are active in defense against parasitic worms and in allergic reactions. Basophils are important in the inflammatory

reaction to injury and release inflammatory molecules when stimulated.

Lymphocytes make up 20% to 25% of the blood leukocyte population. They have large spherical nuclei and are divided into two groups (B and T). B lymphocytes are activated by cell markers called antigens to produce proteins called antibodies (humoral immune response). T lymphocytes are part of the cell-mediated arm of the immune system and destroy abnormal body cells, e.g., cancer, and virally infected cells. Monocytes can leave the circulation to become macrophages, a type of cell that ingests dead and dying cells and bacteria.

Platelets are fragments of cells that play a key role in hemostasis (see pp. 392–393).

▶ **CELLS OF THE BLOOD**
Cells of the blood include red blood cells (upper left), which contain the protein hemoglobin (lower left). About 1% of blood cells are white blood cells, which come in a variety of types.

RED BLOOD CELL (ERYTHROCYTE)

MONOCYTE
Monocytes circulate in the blood for one to two days before entering the body tissues to become macrophages.

MACROPHAGE
Macrophages fight infection by engulfing foreign organisms and debris.

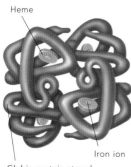

Heme

Iron ion

Globin protein strand

STRUCTURE OF HEMOGLOBIN

NEUTROPHIL
The frontline defense against bacterial invasions, neutrophils engulf and destroy microorganisms.

BASOPHIL
These cells release substances that increase the body's response to invading allergens.

EOSINOPHIL
Eosinophils release enzymes that cause allergic reactions and kill some parasites.

LYMPHOCYTE
There are three types of lymphocyte: natural killer cells and T cells attack foreign invaders directly, and B cells make antibodies.

Plasma proteins

THE PROTEINS THAT ARE SUSPENDED IN THE PLASMA ARE CALLED PLASMA PROTEINS. They include albumin, gamma globulins, transport proteins like transferrin, and clotting factors (fibrinogen). Albumin is made in the liver and is a large protein that cannot normally leave the circulation. It maintains the blood's colloid osmotic pressure (see pp. 378–379), tending to draw water into the circulation from the extracellular spaces. Failure of albumin production, e.g., in chronic liver disease, causes swelling of tissue spaces and accumulation of fluid in body cavities.

Immune proteins are made by leukocytes—principally B lymphocytes and their derivatives called plasma cells. Immune proteins or antibodies (immunoglobulins) such as IgG, IgA, and IgM bind to antigens on foreign invaders, e.g., bacteria, viruses, parasites, and their toxic products, to cause clumping, activate phagocytes to engulf them, or neutralize their toxins.

Transport proteins, e.g., lipoproteins, carry mainly hydrophobic molecules like lipids, i.e., fats, and lipid-soluble substances such as vitamins A, K, and D, around the circulation. If they did not bind to these transport proteins, lipids would tend to clump and obstruct the circulation of blood. Other transport proteins carry essential minerals like iron (bound to transferrin) and copper (bound to ceruloplasmin) around the circulation.

Clotting factors are proteins that engage in a biochemical cascade that converts the blood into a solid gel when there is damage to the vessel wall. A major clotting protein is fibrinogen, which is converted to fibrin when the blood clots.

▶ **MAJOR PLASMA PROTEINS**

This illustration shows the plasma proteins that make up 90% of the total mass (lower pie chart) and plasma proteins that make up 10% (upper pie chart). Albumin makes up the bulk of the plasma proteins (about half), followed by immunoglobulins (immune protein molecules) such as IgG, IgA, IgM, and a variety of complement molecules. The rest of the plasma proteins are clotting factors, e.g., fibrinogen, and transport molecules such as lipoproteins and transferrin. A host of other plasma proteins make up 10% of the total. These have been shown in the upper pie chart, although there is still 1% of the total that consists of other proteins.

MAJOR PLASMA PROTEINS
99% of plasma protein mass

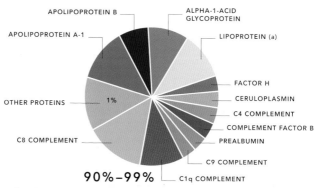

APOLIPOPROTEIN B — ALPHA-1-ACID GLYCOPROTEIN
APOLIPOPROTEIN A-1 — LIPOPROTEIN (a)
FACTOR H
CERULOPLASMIN
C4 COMPLEMENT
COMPLEMENT FACTOR B
OTHER PROTEINS — 1%
PREALBUMIN
C8 COMPLEMENT — C9 COMPLEMENT
C1q COMPLEMENT

90%–99%

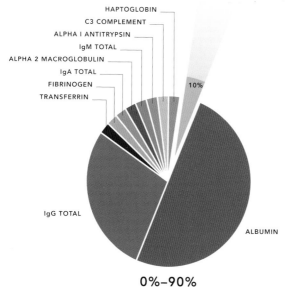

HAPTOGLOBIN
C3 COMPLEMENT
ALPHA I ANTITRYPSIN
IgM TOTAL
ALPHA 2 MACROGLOBULIN
IgA TOTAL
FIBRINOGEN
TRANSFERRIN

10%

IgG TOTAL
ALBUMIN

0%–90%

Function of red blood cells

THE STRUCTURE OF RED BLOOD CELLS (erythrocytes) is optimized for gas transport. The biconcave disk shape maximizes the available surface area for exchange of gases with the surrounding blood, and the cells nucleus-free cytoplasm is packed with hemoglobin (Hb) molecules to carry oxygen and carbon dioxide.

Hemoglobin consists of four globin (protein) chains—two alpha and two beta chains in normal adults—with each holding a heme porphyrin ring molecule with a central iron atom. Each iron atom can bind to a single oxygen molecule to form oxyhemoglobin (HbO_2) when oxygen levels are high, e.g., in pulmonary capillaries. When oxygen levels are low and carbon dioxide levels are high, e.g., at the venous end of the capillary bed, carbon dioxide can bind to Hb to form carbaminohemoglobin, but only 2% of carbon dioxide is carried in this way.

Red blood cells also contain carbonic anhydrase, an enzyme that catalyzes the conversion of carbon dioxide and water to bicarbonate and hydrogen ions. This provides a means of transporting carbon dioxide around the blood as well as buffering to maintain the optimal pH for cellular processes.

▶ **RED BLOOD CELL STRUCTURE AND FUNCTION**

Red blood cells (erythrocytes) are biconcave discs without nuclei (opposite below left). Red blood cells are packed with hemoglobin (opposite below right), which is made up of four globin chains, each with a central heme group and iron atom. Hemoglobin releases oxygen optimally when the red blood cells are passing through systemic capillaries (see oxygen dissociation curve). Aging or defective red blood cells are removed by the spleen (below).

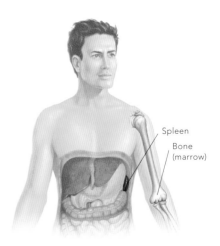

Spleen

Bone (marrow)

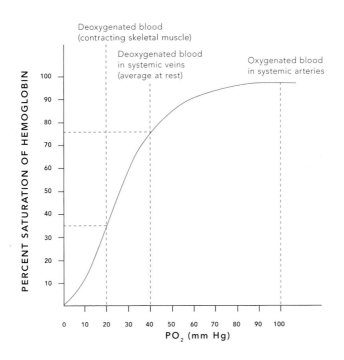

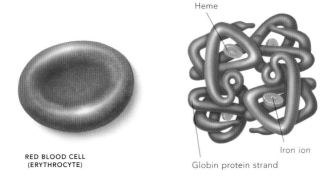

**RED BLOOD CELL
(ERYTHROCYTE)**

Control of red blood cell production

ERYTHROCYTES ARE MADE IN THE RED BONE MARROW OF LONG BONES IN A PROCESS CALLED ERYTHROPOIESIS (red cell formation) that takes between five and seven days and is regulated by the hormone erythropoietin. Erythropoiesis is part of a larger process called hematopoiesis (literally "blood formation"). Once in the circulation, red blood cells last approximately 120 days.

Erythropoiesis is regulated by a negative feedback loop that keeps the hematocrit within optimal levels: too high and blood viscosity interferes with blood flow; too low and oxygen carriage is compromised. The kidneys produce erythropoietin when low oxygen levels are detected in the blood. Erythropoietin then stimulates the red bone marrow to increase the rate of red blood cell production and decrease the time it takes for red blood cells to mature. Once oxygen levels rise, erythropoietin production is turned down.

Red blood cells are made by maturation of a hematopoietic stem cell in the marrow through proerythroblast, erythroblast, and reticulocyte stages. During this process, hemoglobin (Hb) is synthesized and the nucleus and other organelles shrink and are expelled. If red blood cells are mobilized into the circulation rapidly, e.g., after blood loss or in anemia, a significant proportion of circulating red blood cells are still at the reticulocyte stage, i.e., they have residues of the nucleus still present. If iron is in short supply, i.e., iron-deficiency anemia due to chronic blood loss, the red blood cells may be small and pale due to reduced Hb concentration (microcytic and hypochromic red blood cells).

▶ **PRODUCTION OF RED BLOOD CELLS**

Red blood cells are produced in the red bone marrow from a hematopoietic stem cell. The process is regulated by the hormone erythropoietin, which is produced in the kidney. Differentiating red blood cells pass through proerythroblast, erythroblast, and reticulocyte stages.

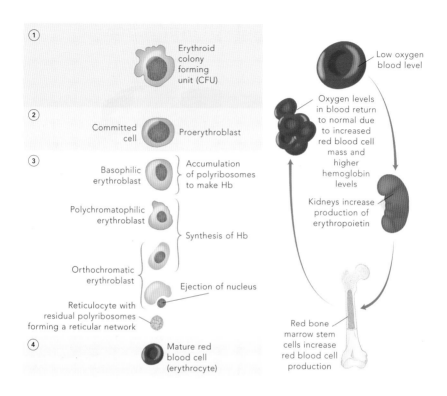

① Erythroid colony forming unit (CFU)

Low oxygen blood level

② Committed cell — Proerythroblast

Oxygen levels in blood return to normal due to increased red blood cell mass and higher hemoglobin levels

③ Basophilic erythroblast

Accumulation of polyribosomes to make Hb

Polychromatophilic erythroblast

Kidneys increase production of erythropoietin

Synthesis of Hb

Orthochromatic erythroblast

Ejection of nucleus

Reticulocyte with residual polyribosomes forming a reticular network

Red bone marrow stem cells increase red blood cell production

④ Mature red blood cell (erythrocyte)

ANEMIA

Anemia is a condition of insufficient hemoglobin in the blood. Normal Hb levels are 13.8 to 18.0 g/dL in males and 12.1 to 15.1 g/dL in females. Common causes of anemia include chronic blood loss, e.g., from chronically heavy menstrual flow or a bleeding cancer, which causes an iron-deficiency anemia (hypochromic microcytic red blood cells). Deficiency of vitamin B12 and folic acid, e.g., due to problems with the stomach or small intestine, causes a megaloblastic (very large red cell) anemia due to problems with DNA synthesis in the hematopoietic cell line.

Blood groups

RED BLOOD CELLS CARRY SURFACE
MOLECULES OR ANTIGENS THAT DIFFER
BETWEEN INDIVIDUALS. If blood from
one individual is transfused into
another person with different red
blood cell antigens, i.e., a different
blood group, a fatal transfusion
reaction may result.

There are several systems of blood
grouping. The ABO system is the most
important and consists of four possible
types—type A, type B, type AB, and
type O. These four ABO system blood
types are based on the presence or
absence of two antigens—A antigen
and B antigen—on the red blood cell
surface. There can be just A, just B,
both A and B, or neither. Individuals
with one particular antigen on their
red blood cells can receive blood of the
same group but will mount an immune
transfusion reaction when they receive
blood from someone with a different
antigen on the red blood cells.

The rhesus system is also important
around the time of childbirth.
Individuals may have the rhesus factor
on their red blood cells (Rh+) or not
(Rh-). If a fetus is Rh+ (from a gene
inherited from the father) and its blood
enters the maternal circulation during
parturition, an Rh- mother may
become immunized to all future fetuses
who are Rh+. These future fetuses will
be attacked by maternal antibodies
that cross the placenta into the fetal
circulation, causing a condition known
as hemolytic disease of the newborn,
or erythroblastosis fetalis. Giving
an Rh- mother an injection of an
immunoglobulin that mops up Rh+
red blood cells at the time of birth,
before they cause immunization, can
prevent future problems.

▶ **ABO BLOOD GROUPS**

There are several systems of blood
antigens that must be considered in
blood typing, but the most important is
the ABO system. Individuals may have
A or B antigens on their red blood cells
(type A or B), both (type AB), or neither
(type O). Someone from one blood
group will make antibodies against
red blood cells from someone with a
different blood group antigen. If an
individual belongs to one group, they
can have blood from the same group or
type O. Type AB individuals can receive
blood from type A, B, or O individuals.

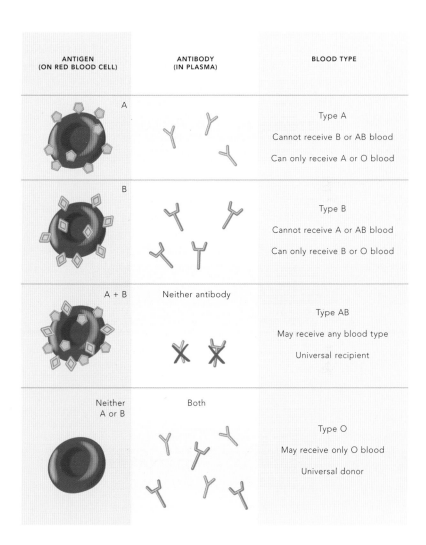

ANTIGEN (ON RED BLOOD CELL)	ANTIBODY (IN PLASMA)	BLOOD TYPE
A		Type A Cannot receive B or AB blood Can only receive A or O blood
B		Type B Cannot receive A or AB blood Can only receive B or O blood
A + B	Neither antibody	Type AB May receive any blood type Universal recipient
Neither A or B	Both	Type O May receive only O blood Universal donor

Hemostasis

HEMOSTASIS MEANS "BLOOD-STOPPING" AND IS THE PROCESS BY WHICH THE BODY PLUGS HOLES IN BROKEN VESSELS. A series of events contributes to this when vascular damage is detected.

First, there is vascular spasm, which is the constriction of arterioles supplying the area in which vessels have been damaged, reducing blood flow to the area and hence blood loss. The second mechanism is the formation of a platelet plug at the site of vessel damage. Platelets become sticky when collagen is exposed, and the endothelial cells of injured blood vessels release a glycoprotein called von Willebrand factor. The sticky platelets plug any holes in the circulation but can also clump abnormally to damaged endothelium in the disease atherosclerosis. Coagulation of the blood is the third part of hemostasis. This depends on the clotting factor cascades that ultimately convert fibrinogen to fibrin and bind the blood elements into a solid gel.

Clotting factors are made in the liver, and coagulation can proceed by two cascades: the intrinsic and extrinsic pathways. The intrinsic pathway is so called because all the required factors are already present within the blood, and it is triggered when the blood comes into contact with exposed collagen. The extrinsic pathway depends on factors outside the blood, e.g., the tissue factors produced by cells outside the endothelium of blood vessels that are exposed when a vessel wall is broken.

Other events in hemostasis are clot retraction, in which actin and myosin in platelets pull the edges of wounds together and squeeze the liquid component out as serum. The body also requires a mechanism to remove coagulated blood once vascular repair has been achieved. This is called thrombolysis and depends on the degradation of fibrin by an enzyme called plasmin.

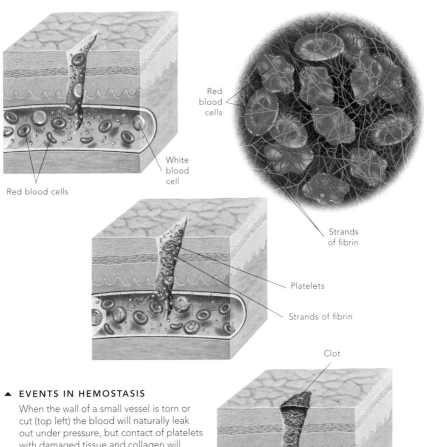

Red blood cells

White blood cell

Red blood cells

Red blood cells

Strands of fibrin

Platelets

Strands of fibrin

Clot

▲ EVENTS IN HEMOSTASIS

When the wall of a small vessel is torn or cut (top left) the blood will naturally leak out under pressure, but contact of platelets with damaged tissue and collagen will trigger aggregation of platelets to form a plug (middle). Coagulation of the extruded blood will also occur as the extrinsic pathway of coagulation is triggered, forming a clot (bottom). The end result is that red blood cells are trapped in a dense network of fibrin protein strands (top right).

Production of clotting factors

CLOTTING FACTORS ARE NAMED ACCORDING TO A SYSTEM OF ROMAN NUMERALS, although they are numbered in order of their discovery rather than the position they come in the coagulation cascade. Most clotting factors are enzymes that are produced in the liver and which then circulate in the blood in their inactive form. The only exceptions to this are factor III, which is a glycoprotein present on the plasma membranes of cells outside the vessels, and factor IV, which is calcium ions already present in the blood.

The process of blood coagulation depends on a biochemical cascade, in which one factor catalyzes the conversion of the next in series, and so on, to the conversion of fibrinogen to fibrin. Four of the clotting factors—II or prothrombin, VII or stable factor, IX or the Christmas factor, and X or the Stuart factor—require the presence of vitamin K for their manufacture. These provide an opportunity for drug-induced modification of blood coagulation by vitamin K analogues like warfarin.

Disseminated intravascular coagulation (DIC) is a condition in which the coagulation mechanisms of the blood are activated inappropriately. DIC can be caused by a variety of conditions, including blood cancers, obstetric conditions, bacterial infection, and burns. Excessive coagulation of the blood can be fatal, such as when blood coagulates to form a thrombus in the small arteries supplying the brain or heart. DIC can lead to multi-organ failure and widespread bleeding because the condition also consumes clotting factors and platelets that may be needed elsewhere in the body.

▶ **SUMMARY DIAGRAM OF COAGULATION CASCADE**

Blood coagulation can be initiated by an extrinsic pathway (due to contact of blood with tissue outside the vessel) or the intrinsic pathway (using factors within the blood). The two pathways use different clotting factors (see flow chart) but have a final common pathway involving thrombin and fibrin formation.

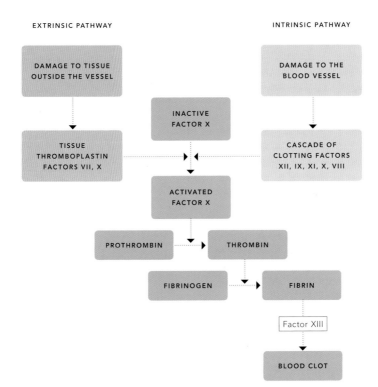

EXTRINSIC PATHWAY

INTRINSIC PATHWAY

DAMAGE TO TISSUE OUTSIDE THE VESSEL

DAMAGE TO THE BLOOD VESSEL

INACTIVE FACTOR X

TISSUE THROMBOPLASTIN FACTORS VII, X

CASCADE OF CLOTTING FACTORS XII, IX, XI, X, VIII

ACTIVATED FACTOR X

PROTHROMBIN → THROMBIN

FIBRINOGEN → FIBRIN

Factor XIII

BLOOD CLOT

HEMOPHILIA

Hemophilia is a condition in which a clotting factor is deficient. Hemophilia A is caused by a shortage of factor VIII, and hemophilia B by inadequate factor IX. Hemophiliacs require transfusions of clotting factors to avoid internal bleeding. This is particularly damaging when it occurs in joints and can lead to serious joint damage.

Types of white blood cells

LEUKOCYTES, OR WHITE BLOOD CELLS, ALL HAVE A PROMINENT NUCLEUS. Some may have granules in their cytoplasm (granulocytes), whereas others do not (agranulocytes). There are three types of granulocytes: neutrophils, eosinophils, and basophils, which are present in the blood at 2,000 to 7,500, 100 to 400, and 20 to 50 cells per mm³, respectively.

Neutrophils have a three- to five-lobed nucleus and granules that stain with neutral dyes. They can cross into the tissues and are attracted in a process called chemotaxis to damaged cells. Neutrophils release their granules at sites of tissue injury to kill bacteria, enhance inflammation, and attract other leukocytes to the area. Neutrophils can also phagocytose (engulf) bacteria and debris.

Eosinophils have a bilobed nucleus and granules in the cytoplasm that stain with eosin. Eosinophils are involved in defense against parasitic worms and in some allergic reactions. Basophils have "S-shaped" nuclei and granules that stain with basic dyes and are released during the inflammatory response.

Lymphocytes are present in the blood at 1,000 to 4,000 cells per mm³. They are divided into types T and B. T lymphocytes are part of the cell-mediated immunity of the body and are active against virally infected and cancerous cells. B lymphocytes transform into plasma cells to produce antibodies against foreign proteins, bacteria, and viruses.

Monocytes are large leukocytes with a "U-shaped" nucleus and are present in the blood at populations of 100 to 700 cells per mm³. They stay in the circulation only a short time before migrating into the tissue spaces, where they become macrophages.

▶ **WHITE BLOOD CELLS**
White blood cells may have intracellular granules (granulocytes like the neutrophil, eosinophil, and basophil) or have a cytoplasm free of granules (agranulocytes like the lymphocyte and monocyte).

MONOCYTE

Monocytes are derived from a stem cell called the granulocyte-macrophage progenitor colony forming unit. Monocytes in the blood have a large indented nucleus and are mobile in response to chemical signals.

MACROPHAGE

Macrophages are monocytes that have migrated from the blood and reside in key organs (lung, spleen, liver, lymph nodes, gut, and bone) where they can engulf foreign material and microorganisms.

NEUTROPHIL

Neutrophils are present in high numbers in the blood and migrate into tissue to engulf bacteria that have been identified by the immune system. Neutrophils can engulf bacteria or expel enzymes onto bacteria in tissue to make pus.

BASOPHIL

Basophils have granules in their cytoplasm that contain heparin and chemicals that control allergic reactions. They play a role in viral infections and chronic inflammatory conditions like rheumatoid arthritis.

EOSINOPHIL

Eosinophils are the main defense cell against parasites. They can also be involved in triggering bronchial asthma.

LYMPHOCYTE

Lymphocytes have a round or slightly indented nucleus with a granule-free cytoplasm. They are divided into T and B types and engage in the humoral and cell-mediated immune responses of the body.

Function of macrophages & granulocytes

MACROPHAGES AND GRANULOCYTES ARE PHAGOCYTIC CELLS, which means that they have the ability to engulf cells and debris. Macrophages and granulocytes belong to the innate limb of the immune system, meaning that they function without previous exposure to a specific antigen. Phagocytosis is achieved by the cell-extending pseudopodia around the object to be engulfed, e.g., a bacterium. The pseudopodia merge, pinching off a vesicle called a phagosome that brings the bacterium into the interior of the cell. The phagosome is then fused with lysosomes containing digestive enzymes and other chemicals, such as hydrogen peroxide and hypochlorous acid, and the bacterium is digested.

Macrophages are activated by several stimuli, including molecules present on the surface of bacteria, debris from dead cells, and signals from other immune system cells. Macrophages are usually the first cells to respond to cellular injury and can also function as antigen-presenting cells, meaning that they can display parts of the pathogens that they have ingested on their cell membrane. T cells can then respond to these and in turn secrete substances that increase macrophage activity.

Neutrophils ingest pathogens, e.g., bacteria, and kill them with hydrogen peroxide, hypochlorous acid, and lysozyme. When pathogens are too big to ingest, the neutrophils release their killing agents onto the invader. When neutrophils are killed and their cell-killing agents are released into tissue spaces, a yellow liquid called pus is produced from the remnants of neutrophils, necrotic tissue, and dead pathogens.

▶ MACROPHAGE AND NEUTROPHIL ACTION

The macrophage and the polymorphonuclear neutrophil granulocyte (neutrophil) are the two main defense cells of the innate immune system. Macrophages engulf foreign material when it is smaller than them but join together to form multinucleated giant cells to wall off foreign material when it is too large to be encompassed by one cell (e.g., retained surgical suture). Neutrophils can also engulf material that is smaller than them (e.g., the meningococcus bacterium) but release the toxic contents of their granules onto invaders when they cannot engulf them. Dead neutrophils and their granular products make up a major component of pus.

MACROPHAGE
Macrophages fight infection by engulfing foreign organisms and debris.

NEUTROPHIL
The frontline defense against bacterial invasions, neutrophils engulf and destroy microorganisms.

Humoral immunity: B lymphocytes & antibody production

HUMORAL IMMUNITY IS THAT ASPECT OF IMMUNE SYSTEM FUNCTION THAT DEPENDS ON SOLUBLE FACTORS LIKE ANTIBODY PROTEINS AND COMPLEMENT TO DESTROY FOREIGN INVADERS. B cells develop in the bone marrow from the lymphoid line of cells. Only about 10% of B cells leave the bone marrow because the B cells that react to self-antigens are destroyed to avoid autoimmune disease. Those B cells that enter the circulation eventually come to reside in the spleen and lymph nodes. If B cells encounter the specific antigen they are intended for, they become activated, dividing repeatedly to produce daughter cells that mature into plasma cells and memory B cells. Plasma cells secrete antibodies immediately against a specific antigen, whereas memory B cells will respond to the antigen on a subsequent exposure.

Antibodies are peptide chains that specifically bind to an antigen on a cell surface or molecule. Five types of antibodies are recognized. IgG is the most prevalent (about 75% to 80%).

It is the only antibody small enough to cross the placenta. IgA is a secretory antibody that is present in secretions from exocrine glands, e.g., saliva, milk, sweat, and mucous membranes. IgM is the largest antibody and consists of five subunits joined in a star shape. It is usually the first antibody type produced when a novel pathogen is encountered. IgE is a type of antibody that binds to parasitic worms and is also implicated in allergic reactions. IgD is located on the surface of B cells and acts as an antigen receptor that helps activate B cells.

Antibodies can destroy pathogens by many mechanisms: they can bind to a group of pathogenic cells and clump them together (agglutination and precipitation); they can bind to the surface of pathogens and stimulate their attack by phagocytes (opsonization); they can neutralize toxins by binding to their active sites (neutralization); they can activate the complement proteins of innate immunity to break pathogens apart (complement activation); or they can stimulate inflammation.

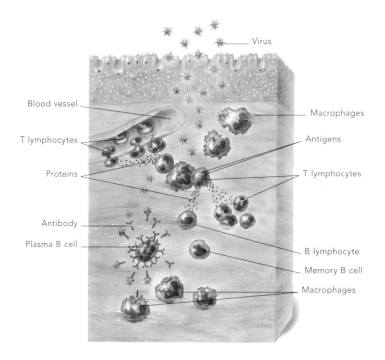

Virus

Blood vessel

T lymphocytes

Proteins

Antibody

Plasma B cell

Macrophages

Antigens

T lymphocytes

B lymphocyte

Memory B cell

Macrophages

▲ THE IMMUNE RESPONSE
TO VIRAL INVASION

This illustration shows the responses of the
immune system to a viral invasion of the
body surface. The epithelium of the body
surface is at the top. The viral antigens
trigger the release of antibodies from
plasma cells (derived from memory B
cells if there has been previous exposure
to this virus). The antibodies bind to viral
particles, which are then engulfed by
macrophages. T lymphocytes are also
mobilized to attack virally infected cells.

Structures involved in immune function: the spleen & lymphoid tissue

THE SPLEEN IS A LYMPHOID ORGAN LOCATED IN THE UPPER LEFT SIDE OF THE ABDOMEN. It has a connective tissue capsule that sends trabeculae into the interior of the spleen. The spleen is divided into red and white pulp. The red pulp is a blood filter that removes old and damaged red blood cells from the circulation. It may also act as a storage site for red blood cells to release them in the event of blood loss. The red pulp also contains macrophages that can engulf bacteria that pass through the red pulp. The white pulp is the immune system component of the spleen and functions much like the lymph nodes, except that the antigens arrive in the spleen by the bloodstream rather than lymph fluid.

MALT and GALT stand for mucosa-associated lymphoid tissue and gut-associated lymphoid tissue, respectively. MALT is the more general term and refers to diffuse and small concentrations of lymphoid tissue in the mucosa of the body in general, e.g., oral cavity, airways, and gut lining.

GALT is a more specific term for those types of MALT found in the gut tube. A good example of GALT would be the Peyer's patches that are found in the lining of the ileum. MALT contains T and B lymphocytes, macrophages, and antibody-producing plasma cells. MALT is in an excellent position to respond to foreign antigens passing through the epithelial surface into the body's tissues and to begin the immune response to those.

▶ **GUT-ASSOCIATED LYMPHOID TISSUE**

The gastrointestinal tract is a major interface between the body and the external environment. Gut-associated lymphoid tissue (GALT) must deal with not only pathogenic micro-organisms and parasites but also the bacteria and yeast cells of the normal gut flora. GALT consists predominantly of nodules of lymphoid tissue embedded in the submucosa of the gut wall.

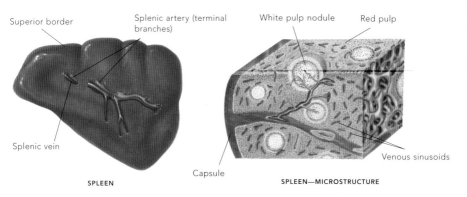

Superior border

Splenic artery (terminal branches)

Splenic vein

SPLEEN

White pulp nodule

Red pulp

Capsule

Venous sinusoids

SPLEEN—MICROSTRUCTURE

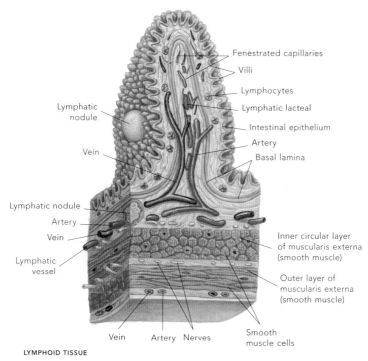

Fenestrated capillaries

Villi

Lymphocytes

Lymphatic lacteal

Intestinal epithelium

Artery

Basal lamina

Lymphatic nodule

Vein

Lymphatic nodule

Artery

Vein

Lymphatic vessel

Inner circular layer of muscularis externa (smooth muscle)

Outer layer of muscularis externa (smooth muscle)

Vein   Artery   Nerves

Smooth muscle cells

LYMPHOID TISSUE

Structures involved in immune function: the thymus & tonsils

THE THYMUS IS A LYMPHOID ORGAN LOCATED IN THE FRONT OF THE CHEST CAVITY. It is most active during early childhood, when it serves the important function of transforming immature T lymphocytes into mature forms that are able to both react to foreign antigen fragments bound to their cell membrane and not respond to self-antigens, i.e., surface markers of the body's own cells, that are similarly bound. Both of these functions are critically important for T lymphocytes to protect the body against disease-causing organisms and to avoid damaging the body's own tissues. The process depends on interactions between T lymphocytes and special epithelial cells in the cortex of the thymus between birth and five years of age.

The tonsils are aggregations of lymphoid tissue at the entrances to the respiratory and gastrointestinal tracts. The tonsils occupy these strategic positions so that they can monitor incoming air, water, and food for the presence of disease-causing micro-organisms and/or foreign proteins and toxins. The nasopharyngeal tonsils are located behind the nasal cavity on the posterior wall of the nasopharynx and on the tubal elevations above the pharyngeal openings of the auditory tubes that connect the middle ear to the nasopharynx. The palatine tonsils are located on the lateral wall of the passage that connects the oral cavity to the oropharynx and sit in a shallow depression called the palatine fossa, between the palatoglossal and palatopharyngeal folds. The lingual tonsil is located on the surface of the posterior one-third of the tongue and curves downward toward the epiglottis.

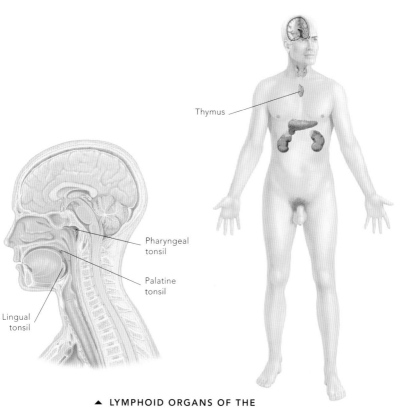

Thymus

Pharyngeal
tonsil

Palatine
tonsil

Lingual
tonsil

▲ LYMPHOID ORGANS OF THE BODY

Lymphoid tissue is mainly disseminated
throughout the body, but discrete
collections at key sites can also be
recognized. These include the tonsils
at the entrances to the gastrointestinal
and respiratory tracts (lingual, palatine,
and nasopharyngeal tonsils) and the
thymus in the chest cavity for production
of T cells during early childhood.

Cellular immunity: T lymphocytes & cell-mediated immunity

CELL-MEDIATED IMMUNITY INVOLVES SEVERAL DIFFERENT CLASSES OF T CELLS. These include the helper T (CD4) cells and the cytotoxic T (CD8) cells. Cell-mediated immunity responds mainly to cells infected with intracellular pathogens such as viruses and intracellular bacteria, e.g., tuberculosis. It is also effective against cancer cells and foreign cells in transplanted organs.

T cells are originally made in the bone marrow, but they move to the thymus gland to mature. Each T cell responds to a particular antigen and can produce a series of identical clones. The thymus screens those T cells that can respond to pathogens and eliminates those that cannot. It is also important that T cells do not damage the body's own tissues. Any T cells that would respond to the body's own cells are eliminated to ensure that T cells released to the circulation are self-tolerant.

Every T cell has a receptor on its cell surface that must bind to a specific antigen before the T cell can be activated, so each T cell will attack a very specific target. Helper T cells have no direct cell-killer effects themselves, but they secrete cytokines that activate other components of the immune system, e.g., macrophages, cytotoxic T cells, and B cells. Failure of helper T cells (as occurs in AIDS) can cause failure of the entire immune response.

Cytotoxic T cells kill other cells. They can detect abnormalities in any cell with a nucleus, so they can eliminate cancer cells, foreign cells (in transplants), and cells infected with intracellular pathogens, e.g., bacteria and viruses. Killer T cells bind to their target and secrete a protein called perforin that punches holes in the cell membrane.

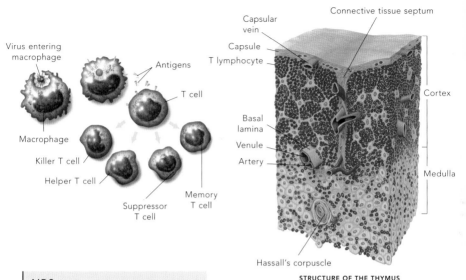

Virus entering macrophage

Antigens

T cell

Macrophage

Killer T cell

Helper T cell

Suppressor T cell

Memory T cell

Capsular vein

Capsule

T lymphocyte

Basal lamina

Venule

Artery

Connective tissue septum

Cortex

Medulla

Hassall's corpuscle

STRUCTURE OF THE THYMUS

AIDS

AIDS stands for acquired immunodeficiency syndrome, which is a condition of impaired cell-mediated immunity due to infection with the Human Immunodeficiency Virus (HIV), usually type 1, subtype B in the west. AIDS starts with a mild flu-like illness two to six weeks after exposure. This is followed by a reduction in the population of the body's CD4 T-helper cells, leading to opportunistic infections, tumors like lymphomas and Kaposi's sarcoma, and a type of dementia when the virus enters the brain.

▲ CELLS AND ORGANS OF CELL-MEDIATED IMMUNITY

The main elements of the cell-mediated immune response are the T lymphocytes. These get their name because they spend some time during their development in the thymus, a lymphoid organ (see right-hand illustration) in the front of the chest cavity. Lymphocytes can be classified into helper T, suppressor T, memory T, and killer T types. Macrophages that have ingested foreign material may present key molecules (antigens) for the T lymphocytes to respond to. T lymphocytes are active against virally infected cells, cancer cells, and foreign tissue, e.g., in a graft or transplant.

The immune system & cancer

CANCER CELLS ARE BODY CELLS THAT HAVE UNDERGONE MUTATIONS THAT CAUSE THEM TO LOSE THEIR DIFFERENTIATION, i.e., they become more primitive or less specialized. Cancer cells are no longer under the body's regulation of its cell populations; cancer cells begin to undergo uncontrolled cell division, and they lose their normal attachments to surrounding tissues. Cancer cells form a malignant tumor that grows uncontrollably and may spread to distant sites in the body (metastasize).

Certain cells of the immune system (T cells and natural killer cells) perform an immune surveillance of the cells in the body for tumor antigens—cell surface markers that indicate that a cell is cancerous. Cancer cells can be destroyed by the immune system in the following way: (i) cancer cells damage the surrounding tissues and cause an inflammatory response, releasing inflammatory markers that are attractive to immune system cells; (ii) natural killer cells (a type of innate immunity cell) migrate to the region and secrete immune proteins like interferon as they begin to kill the cancer cells; (iii) tissue macrophages are attracted by interferon and also activated to secrete tumor necrosis factor, which makes cancer cells undergo apoptosis (cell death); (iv) fragments of dead tumor cells are taken up by dendritic cells in the tissue and then presented as tumor antigens to nearby T helper cells; (v) the T helper cells then activate killer T cells to attack the tumor cells.

▶ **CANCER GROWTH**

Cancer often arises in surface or glandular tissues (epithelia), in which case it is more properly called carcinoma, e.g., carcinoma of the breast, prostate, stomach, colon, kidney, bladder, uterus, cervix, or lung. Cancer of the connective tissues is called sarcoma, e.g., osteosarcoma, chondrosarcoma, or fibrosarcoma. A key problem with cancer cells is their tendency to spread to other parts of the body. This may be via the lymphatics, in which case cancer will be found in the lymph nodes, or by the bloodstream (usually venous channels like the portal vein to the liver or the systemic veins to the lung).

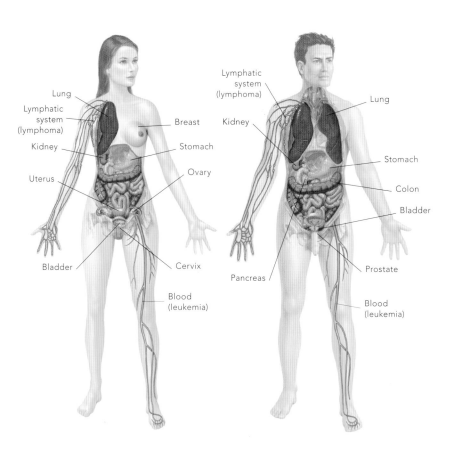

Lung

Lymphatic
system
(lymphoma)

Kidney

Uterus

Bladder

Breast

Stomach

Ovary

Cervix

Blood
(leukemia)

Lymphatic
system
(lymphoma)

Kidney

Lung

Stomach

Colon

Bladder

Pancreas

Prostate

Blood
(leukemia)

Cancer

CANCER IS THE POPULAR TERM FOR ANY MALIGNANT NEOPLASM (invasive new growth). Key features of cancer are the tendency to grow rapidly, invade locally, and spread to distant sites in the body (metastasize). Most cancers arise from an epithelial surface (body surface or glands derived from epithelium) and are called carcinomas. Cancers of the connective tissue, e.g., bone, cartilage, and muscle, are called sarcomas, whereas white blood cancers are called leukemias. Carcinomas may arise from benign tumors (adenomas) of an epithelial surface, e.g., the lining of the bowel.

For a carcinoma to invade adjacent tissues, it must break through the basement membrane upon which the epithelium of origin is located. Cancer cells first reduce their expression of cell adhesion molecules that hold epithelial cells in place. They then release enzymes (collagenase, cathepsins, and hyaluronidase) that break down the fibers of the basement membrane. When cancer cells start their invasive phase, they secrete motility factors to direct the movement of cells and angiogenic factors to encourage the growth of new vessels to the region. The tumor cells can then enter the newly formed vessels, which often have weak walls. A tumor can also enter lymph channels and spread through the lymph nodes of the body. The tumor continues on into the general circulation to reach distant sites (metastasize).

Cancer may cause death by interfering with the normal immune response, e.g., leukemic cells displacing normal immune system cells, leading to death from overwhelming infection. Local spread of tumor may obstruct critical organs, e.g., pelvic tumors may obstruct the ureters, leading to renal failure, or the rectum or bladder, leading to an inability to urinate or defecate. Metastatic spread of a tumor may displace so much of the tissue of vital organs like the liver or kidney that multi-organ failure occurs. Tumor growth also consumes vital nutrients, leading to wasting of normal tissues, and tumors may secrete factors that cause cachexia—a wasting condition in which there is loss of appetite, fatigue, and weakness.

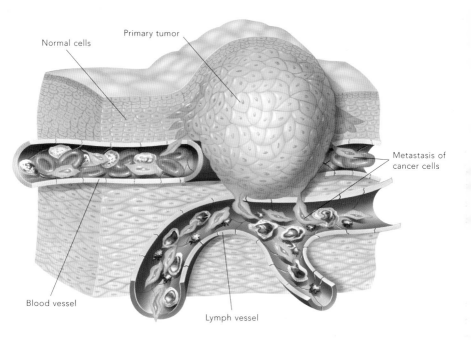

Normal cells

Primary tumor

Metastasis of cancer cells

Blood vessel

Lymph vessel

▲ **METASTASIS BY A MALIGNANT NEOPLASM**

Malignant neoplasms (cancers) can spread locally or to distant sites (metastasize). To reach distant organs, carcinoma cells must break through the basement membrane of the epithelium they arise from and then enter the capillaries and lymphatic channels draining their site of origin. Cancer cells then flow passively with the blood until they reach another capillary bed (lung, liver, brain) where they lodge and grow.

Autoimmune diseases

NORMALLY, THE BODY'S IMMUNE SYSTEM HAS A NATURAL TOLERANCE OF THE BODY'S OWN TISSUE AND CELLS. However, sometimes the self-tolerance mechanisms fail, with the result that populations of self-reactive T and B cells are produced. The B cells produce antibodies (autoantibodies) that bind to antigens on the body's own cells in a phenomenon called autoimmunity. Autoantibodies can spread throughout the body, so autoimmune disease typically involves multiple sites in the body at the same time. The resulting tissue damage can have serious consequences and cause death.

Some types of autoimmune disease, e.g., multiple sclerosis, occur when normal tissue antigens that are usually hidden or sequestered from the immune system, e.g., the myelin coat of nerve cell axons hidden in the central nervous system behind the blood-brain barrier, suddenly become exposed by infection or trauma and excite a reaction by the T cells of the immune system.

Sometimes foreign antigens, e.g., those on streptococcal bacteria, may mimic the antigens on the surface of normal body cells. When the body mounts an immune attack against the bacteria, it also damages its own tissues. This is seen with rheumatic fever, in which autoantibodies attack the heart valves and cardiac muscle.

Some cells may inappropriately express cell surface molecules that excite the immune system. This is seen with juvenile onset diabetes mellitus, in which pancreatic islet beta cells are destroyed by T cells.

Finally, some pathogens can produce inflammatory signaling molecules that non-specifically activate immune system cells to damage body tissues. This is seen in the condition of systemic lupus erythematosus.

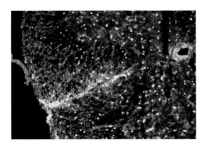

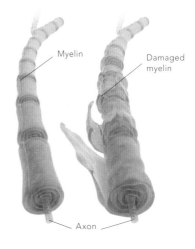

Myelin · Damaged myelin

Axon

NORMAL NERVE **DAMAGED NERVE**

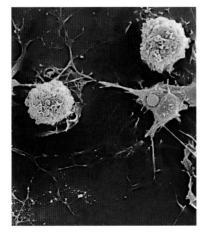

▲ **EXAMPLES OF AUTOIMMUNE DISEASE**

Autoimmune disease occurs when the body's own immune defenses attack the body's tissues. An example is multiple sclerosis. In multiple sclerosis, the body's immune system attacks the myelin sheaths of nerve fibers (top right). The upper left image shows cells of a spinal cord affected by multiple sclerosis. The red fibers are made by supporting cells (glia) of the brain in an attempt to repair damage. The bottom left image shows immune cells of the brain (microglia) attacking and ingesting the cells that make myelin (oligodendrocytes).

End-of-life physiology

WHAT DEFINES DEATH? Death can be defined as a cessation of the biological functions that sustain a living organism. The death of a mammal is traditionally characterized by the cessation of breathing and cardiac arrest, i.e., no pulse, but modern medical interventions can sustain ventilation artificially and restart the heart with electrical stimulation.

In modern medicine, brain death may be used as a definition, i.e., a person is considered brain-dead when there is cessation of brain function as indicated by absence of cerebral cortical activity on the EEG (electroencephalograph). But this still has problems. A person may be brain-dead, i.e., with no electrical activity in the cerebral cortex, but may still be able to sustain circulation and lung ventilation, control body temperature, heal wounds, and even gestate a fetus. Furthermore, some drugs or medical conditions, e.g., hypoglycemia, hypothermia, and hypoxia, may suppress electrical activity. For this reason, to provide a reliable measure, assessment of

brain death by EEG activity must be performed at widely separated points in time under carefully defined conditions.

What happens at death? Ultimately, death is due to a failure of the cardiovascular and respiratory systems, i.e., the heart stops beating and the respiratory muscles stop ventilating the lungs. Close to the time of death, the individual may exhibit Cheyne-Stokes ventilation, in which there is an oscillation in the depth and rate of breathing between hyperventilation and absence of breathing. This is due to defective feedback control of lung ventilation.

After the circulation stops, the skin becomes pale (pallor mortis) between 15 minutes and two hours after death. Blood starts to settle in the lower parts of the body (livor mortis), and the body temperature starts to drop (algor mortis). The limbs of the corpse become stiff (rigor mortis) as the actin and myosin filaments in the skeletal muscle cross-link and are unable to detach because the adenosine triphosphate (ATP) has been used up.

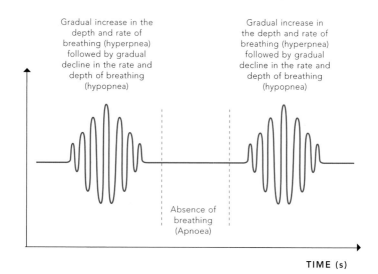

Gradual increase in the depth and rate of breathing (hyperpnea) followed by gradual decline in the rate and depth of breathing (hypopnea)

Gradual increase in the depth and rate of breathing (hyperpnea) followed by gradual decline in the rate and depth of breathing (hypopnea)

Absence of breathing (Apnoea)

DEPTH OF VENTILATION

TIME (s)

▲ **CHEYNE-STOKES VENTILATION**

If blood flow to respiratory centers in the brain is sluggish, or those centers are poorly responsive to stimuli (both of which happen toward the end of life), then the patient may exhibit a type of breathing where there is an oscillation between deep ventilation of the lungs (hyperpnea) and no lung ventilation at all (apnea).

Normal ranges & values

NERVOUS AND SENSORY SYSTEMS

Conduction velocity of large myelinated fiber: **80 to 120 m/s**

Conduction velocity of small unmyelinated fiber: **0.5 to 2 m/s**

Spinal segments supplying the diaphragm: **C3 to C5**

Spinal segments supplying the upper limb: **C5 to T1**

Spinal segments supplying the sympathetic outflow: **T1 to L1**

Spinal segments supplying the lower limb: **L2 to S3**

Spinal segments supplying the bowel and bladder: **S2 to S4**

Parasympathetic outflow: **cranial nerves 3, 7, 9, and 10; sacral segments 2 to 4**

Near point of eye (closest distance that can be focused on): **100 mm at age 25**

Frequency range of normal human hearing: **20 Hz to 20,000 Hz**

Decibel levels (normal conversation): **60 to 70 dB**

CARDIOVASCULAR SYSTEM

Heat rate at rest: **60 to 70 beats per minute**

Systolic blood pressure at rest: **120 mm Hg**

Diastolic blood pressure at rest: **80 mm Hg**

Total blood volume: **5 liters**

Stroke volume: **70 mL**

RESPIRATORY SYSTEM

Normal respiration rate: **12 breaths per minute**

Tidal volume: **500 mL**

Anatomical dead space: **150 mL**

pO_2 air: **160 mm Hg**

pCO_2 air: **0.3 mm Hg**

pO_2 alveoli: **105 mm Hg**

pCO_2 alveoli: **36 mm Hg**

GASTROINTESTINAL SYSTEM
pH of stomach juices: **1.5 to 3.5**
Transit time of gastrointestinal tract: **18 hours to 3 days**

URINARY SYSTEM
Glomerular filtration rate: **125 ml/min**
Tubular reabsorption rate: **124 ml/min**
Urine output: **1 ml/min or approximately 1.5 L/day**

COMPOSITION OF URINE
Osmotic concentration: **850 to 1340 mOsm/L**
Specific gravity: **1.003 to 1.030**
pH: **4.5 to 8.0, with a mean of 6.0**
Bacterial content: **Nil, urine should be sterile**
Red blood cells: **100/mL**
White blood cells: **500/mL**
Sodium: **330 mg/dL**
Potassium: **166 mg/dL**
Chloride: **530 mg/dL**
Calcium: **17 mg/dL**
Urea: **1.8 g/dL**
Creatinine: **150 mg/dL**
Ammonia: **60 mg/dL**
Uric acid: **40 mg/dL**
Urobilin (yellow pigment): **125 µg/dL**

Normal ranges & values *Continued*

BLOOD

BLOOD CHEMISTRY

pO_2 of systemic arterial blood: **75 to 100 mm Hg**
pCO_2 of systemic arterial blood: **35 to 45 mm Hg**
Sodium: **138 mM**
Potassium: **4.4 mM**
Chloride: **106 mM**
Bicarbonate: **27 mM**
pH: **7.35 to 7.45**
Urea: **10 to 20 mg/dL**
Creatinine: **1 to 1.5 mg/dL**
Ammonia: **< 0.1 mg/dL**
Albumin: **3.6 to 4.7 g/dL**
Glucose (whole blood, fasting): **3.3 to 5.6 mM (60 to 100 mg/dL)**
Hemoglobin: **male: 13.8 to 18.0 g/dL; female: 12.1 to 15.1 g/dL**
Hematocrit (proportion): **male: 0.38 to 0.54; female: 0.35 to 0.48**

CELLULAR COMPOSITION

Red blood cell mean volume: **80 to 100 fL (femtoliters 10^{-15} liters)**
Red blood cell mean Hb concentration: **310 to 360 g/L**
Total leukocyte population: **4.0 to 11.0 x 10^9/L**
Neutrophil population: **2.0 to 7.5 x 10^9/L**
Lymphocyte population: **1.0 to 4.0 x 10^9/L**
Monocytes population: **0 to 1.0 x 10^9/L**
Eosinophil population: **0 to 0.5 x 10^9/L**
Basophil population: **0 to 0.3 x 10^9/L**
Platelet population: **150 to 450 x 10^9/L**

COMPOSITION OF SEMEN

Volume of ejaculate: **2 to 5 ml**

Concentration of spermatozoa: **20 million to 100 million per ml**

pH: **7.2 to 7.7**

Percentage of motile forms: **more than 60%**

Percentage of abnormal forms: **less than 40%**

Fructose: **224 mg/dL**

IMPORTANT NOTE: THESE VALUES ARE PROVIDED FOR EDUCATIONAL PURPOSES ONLY AND SHOULD NOT BE USED FOR SELF-DIAGNOSIS. SEE YOUR HEALTH CARE PROFESSIONAL IF YOU HAVE CONCERNS WITH PERSONAL TEST RESULTS. REFERENCE NORMAL VALUES MAY VARY BETWEEN TESTING LABORATORIES DEPENDING ON THE SAMPLE REFERENCE POPULATION.

Units of measurement

UNITS OF LENGTH

m – meter. 1 m is equivalent to 1.09 yards, 3.28 feet, or 39.4 inches.

mm – millimeter – 10^{-3} meter. 1 mm is equivalent to 0.0394 inch.

µm – micrometer – 10^{-6} meter

nm – nanometer – 10^{-9} meter

UNITS OF VOLUME

mL – milliliter. 1 mL is equivalent to 0.0338 US fluid ounce.

mm^3 – cubic millimeter. 1 cubic millimeter is equivalent to 0.0000338 US fluid ounce.

fL – femtoliter – 10^{-15} liter or one-quadrillionth of a liter.

UNITS OF MASS

g – gram. 1 g is equivalent to 0.0353 ounce.

mg – milligram. 1 mg is 0.001 g.

UNITS OF NUMBER OF ATOMS AND CONCENTRATION

mol – mole. 1 mole of a pure substance contains 6.022×10^{23} atoms.

M – molar. This is a unit of concentration, so a 1M solution contains 1 mol per liter.

mM – millimolar. A 1 mM solution contains 0.001 mol per liter.

mOsmM – milliosmolar. This is a term related to mM, but refers to the number of atoms or molecules of a dissolved substance that contribute to the osmotic pressure of a solution.

UNITS OF TEMPERATURE

°C – degrees Celsius. Normal body cavity temperature is 37.0°C or 98.6°F (Fahrenheit scale).

UNIT OF PRESSURE

mm Hg – mm mercury. A pressure of 10 mm Hg is sufficient to support a column of mercury 10 mm high.

UNIT OF VOLTAGE

mV – millivolt – 10^{-3} volts. The volt is a unit of electrical potential difference and electromotive force.

UNIT OF FREQUENCY

Hz – Hertz. This is a unit of the number of cycles per second for an oscillating phenomenon. 1 Hz is one cycle per second.

UNIT OF SOUND INTENSITY

dB – decibel. 10 dB is the equivalent of a 10-fold increase in the loudness of a sound.

UNITS OF ENERGY

J – Joule. 1 J is equivalent to 0.239 calories. One apple contains approximately 523 kJ (523 thousand joules) of chemical energy.

Important note: Dieticians use a term Calorie (note capital "C"), which is 1,000 calories. So the usual daily dietary energy requirement for an adult male is 2,600 Calories, or 2,600 kcal, or 10,900 kJ.

Index

Acknowledgments

The publisher wishes to thank Dr Derek Scott, Senior Lecturer in Integrative Physiology & Pharmacology at the Institute of Education in Medical & Dental Sciences, University of Aberdeen, for his invaluable help during the production of this book.

Picture Credits